NEW

Co-ordinat

SCIENCE

D0714922

Chemistry

second edition

RoseMarie Gallagher   Paul Ingram   Peter Whitehead

Oxford University Press

3300475106

Oxford University Press, Walton Street, Oxford OX2 6DP

Oxford New York
Athens Auckland Bangkok Bogota Bombay
Buenos Aires Calcutta Cape Town Dar es Salaam Delhi
Florence Hong Kong Istanbul Karachi
Kuala Lumur Madras Madrid Melbourne
Mexico City Nairobi Paris Singapore
Taipei Tokyo Toronto

and associated companies in
Berlin Ibadan

Oxford is a trade mark of Oxford University Press

First Published 1989 (ISBN 0 19 914316 1)
Reprinted 1990, 1991, 1992, 1993, 1994

2nd edition published 1996. Reprinted 1996

ISBN 0 19 9146527

Typeset in 10/12pt Palatino by Acc Computing
Printed in Spain
by Gráficas Estella

Acknowledgements

*The publisher would like to thank the following for their kind
permission to reproduce the following photographs:*

p6 Biophoto (bottom centre and right), C James Webb (top
centre), **p9** Dominic Photography/C Ashmore, **p11** Zefa
Photo Library, **p20** Fisons, **p26** De Beers (top centre), Derby
Museum and Art Gallery (bottom right), **p27** Zefa, **p28**
Hunting Films (top right), Hulton Deutsch (bottom right),
p32 Zefa (bottom right), **p33** Science Photo Library/E Willard
Spurr/US Library of Congress, **p34** Zefa (top), Britain on
View/Barry Hicks (bottom), **p39** SPL/Dr J Burgess (left),
J Allan Cash (right), **p41** SPL/C Priest (top), AEA Harwell
(bottom), **p42** AEA Harwell (top and centre right), SPL/
S Stammers (bottom), **p43** Barnabys, **p46** BOC Ltd, **p48**
P Brierley (centre), **p56** University of Liverpool (top centre),
Lead Sheeting Association (bottom right), **p57** De Beers
(top right), **p60** Lead Sheeting Association (top left), SPL/
V Fleming (top centre), Hutchinson Picture Library (top right),
SPL/NASA (bottom right), **p61** Westland Aerospace (top
right), LTG Laboratories (bottom left), SPL (bottom centre),
Tesco (bottom right), **p62** Pilkington Glass (top left), SPL/

P Ryan (top right), Worcester Pottery (bottom left and right),
p63 Wagner Public Relations, Worcester Pottery (top right),
SPL/S Patel (centre right), **p66** J Allan Cash (top right),
Courtaulds (left), Robert Harding Picture Library (right),
p67 SPL/J Hinsch (top right), SPL/A Syred (right), **p68**
Allsport, British Cement Association (bottom centre and
right), **p69** SPL/A and H Frieder Michler (centre top), J Allan
Cash (right), **p72** Hulton Deutsch Picture Library, **p82** SPL/
S Stammers, **p83** Zefa (top right), **p84** Aga (top), **p88** Zefa,
p89 Crown Paints, **p90** Barnabys (centre right), **p91** Alfa
Romeo (top right), Baxi Heating (centre right), Chubb Fire Ltd
(bottom right), **p93** J Allan Cash (bottom right), **p98** Black
and Decker, **p99** Central Electricity Board, **p106** ICI, **p107**
Severn Trent Water (left), ICI (right), **p108 and 109** ICI,
p112 Barnabys (bottom right), British Steel (top right), Zefa
(centre top), **p114** Hutchison Picture Library, **p124** SPL/
A and H Frieder Michler (bottom right), **p125** SPL/
Dr J Burgess, **p135** NHPA/S Dalton (top), Hutchison Picture
Library (right), **p139** Body Shop (bottom right), **p142**
Prestige (right), J Allan Cash (left), **p143** Barnabys (top left),
p146 Thermit Welding, **p148** Royal Mint, **p150** Barnabys
(top right), Garrard and Company (bottom), **p151** Barnabys
(centre right), Geological Museum (top left, centre and top
right), **p152** Barnabys, **p153** Safeway, **p154** J Allan Cash
(centre bottom), Unigate (top left), **p156** Alcan Aluminium,
Jonquiere Analytical Centre (top left), NASA/SPL (bottom
right), **p157** British Alcan (top right), **p158** Zefa (right),
Camera Press (top right), **p160** J Allan Cash (centre and
bottom right), **p161** British Alcan (bottom right), Heinz (top
right), **p164** Barnabys (top), Camera Press (bottom), **p165**
BOC Ltd, **p166 and 167** ICI, **p168** Holt Studios (left), Oxford
Scientific Films/R Toms (centre), SPL/A Hart-David (right), ICI
(bottom), **p169** Farmers Weekly (left), Holt Studios (top right),
R Harding (right), **p171** Pictor International (centre right),
p172 SPL/M Dohrn (bottom left), Supersport (bottom right),
OSF (bottom centre), **p173** British Gas (left and centre), S and
R Greenhill (right), **p176** GeoScience Features (bottom
centre), Sulphur Institute (left and right), **p177** Deckers (left),
Friends of the Earth (top right), **p178** ICI, **p180** Hulton
Deutsch, **p184** De Beers (top), Barnabys (centre), Michael
Holford (bottom), **p186** SPL (centre right), Barnabys (top
right), Holt Studios (top left and centre), **p185** Holt Studios,
p186 Camera Press, **p190** Zefa (bottom and top), **p191** S and
R Greenhill (bottom), **p192** British Petroleum (left and bottom
right), Thames Water Authority (centre), Colorsport (right),
p193 J Allan Cash, **p194** British Petroleum (top), Esso
(bottom), **p197** SPL/P Menzel, **p198** J Allan Cash (top),
Barnabys (bottom), **p199** UCE School of Jewellery, **p202**
J Allan Cash (top), **p203** GeoScience Features (top left),
p204 Hulton Deutsch, **p206** SPL/NASA (both), **p209** SPL
(left), **p210** BOC Ltd, **p211** BOC Ltd (all), **p212** J Allan Cash
(top), SPLA/A Dowsett (bottom), **p213 and 214** R Harding,
p215 SPL, **p217** R Harding, Permutit (bottom), **p225** Zefa,
p231 SPL, GeoScience Features (bottom), **p234** GeoScience
Features, **p238** British Geological Survey, **p241 and 243**
GeoScience Features.

Additional photography by Chris Honeywell, Peter Gould,
and Tony Waltham.

The illustrations are by:
Harriet Dell, Michael Eaton, Jeff Edwards, Nick Hawken,
Illustra Graphics, Peter Joyce, Ben Manchipp, Mark Oliver,
Tony Simpson, Julie Tolliday, and Galina Zolfaghari.

Introduction

Science is about asking questions. Chemistry is the science that asks questions about materials, the differences between them, how they react with one another and how heat or other forms of energy affect them. What is water made of? What happens when hydrogen burns? How are plastics made? All these questions are of interest to chemists.

Many experiments have been done to find the answers to questions like those above. This book will tell you some of the answers. It explains the facts about chemistry that you will need to know when studying chemistry as part of a GCSE Co-ordinated Science course, or as a single GCSE subject.

Everything in this book has been organized to help you find out things quickly and easily. It is written in two-page units. Each unit is about a topic you are likely to study. The units are grouped into sections.

● Use the contents page
If you are looking for information on a large topic, look it up in the contents list. But if you cannot see the topic you want, then . . .

● Use the index
If there is something small you want to check on, look up the most likely word in the index. The index gives the page number where you'll find information about that word.

● Use the questions
Asking questions and answering them is a very good way of learning. There are questions at the end of every unit. At the end of each section, you can test yourself using exam-level questions. The answers to numerical questions are at the back of the book.

Chemistry is an important and exciting subject. We hope that this book helps you with your studies, that you enjoy using it, and that at the end of your course, you agree with us!

RoseMarie Gallagher
Paul Ingram
Peter Whitehead   January 1996

Contents

1.1 Everything is made of particles

Everything around you is made up of very tiny pieces, or **particles**. Your body is made of particles. So is your desk, your chair, and this book.

A penny is made up of about 34 000 000 000 000 000 000 000 particles.

A small raindrop contains about 1 000 000 000 000 000 000 000 particles.

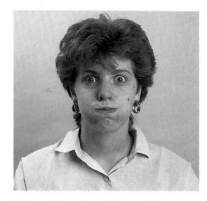

Take a deep breath — and you will breathe in about 40 000 000 000 000 000 000 000 particles from the air.

These particles are so tiny that it is impossible to pick up just one of them, and look at it. However, there is a machine that is powerful enough to take pictures of *groups* of particles. It is called a **scanning electron microscope**. Below are some photographs of a needle taken with this machine.

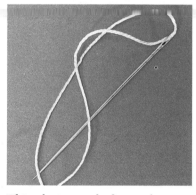

This photograph shows the needle as you would see it.

This is the eye of the needle, magnified about 130 times. You can see that the surface is not perfectly smooth.

This is part of the eye, magnified 60 000 times. The photograph shows that it is made up of clusters of particles.

The scanning electron microscope is very powerful, but still not powerful enough to show just one particle. Each shape in the last picture above contains millions of smaller particles.
Your school laboratory does not have a scanning electron microscope. However, even without one, you can still find evidence that things are made of particles. Some examples are given on the next page.

A crystal dissolving

When a crystal of potassium manganate(VII) is placed in a beaker of water, the water slowly turns purple.

Explanation Both the crystal and water are made of particles. The colour spreads because purple particles leave the crystal and mix with water particles. The crystal **dissolves**.

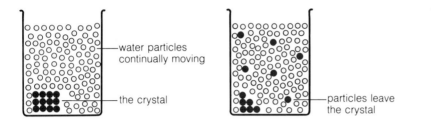

water particles continually moving

the crystal

particles leave the crystal

because the particles are all moving they become evenly mixed

This movement of different particles among each other, so that they become evenly mixed, is called **diffusion**.

Two gases mixing

Air is colourless. Bromine vapour is a red-brown colour and is heavier than air. When a jar of air is placed upside-down on a jar of bromine vapour, the red-brown colour spreads up into it. After a few minutes, the gas in both jars looks the same.

Explanation Both air and bromine are made of tiny moving particles. These collide with each other and bounce about in all directions, so that they become evenly mixed.
This is another example of diffusion.

Before and after: the diffusion of bromine in air.

The movement of smoke

Smoke can be examined by trapping some in a small glass box, shining a light through it sideways, and looking at it with a microscope. The smoke specks show up as bright shiny spots that dance around jerkily. They are never still.

Explanation The smoke specks move because they are knocked about by moving particles of air. (But you can't see the air particles because they are too small.)
In the same way, pollen dances about on the surface of water, because it is bombarded by tiny moving water particles. The warmer the water, the faster the pollen moves. The movement of smoke and pollen was first discovered by a scientist named Robert Brown, about 150 years ago, so it is called **Brownian motion**.

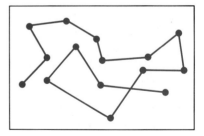

This shows the path of a single smoke speck. It changes direction because it is bombarded by air particles.

Questions

1 Why does the purple colour spread, when a crystal of potassium manganate(VII) is placed in water?
2 What is diffusion? Explain how bromine vapour diffuses into air.

3 What is Brownian motion? Sketch the movement of a particle during Brownian motion.
4 On a sunny day, you can sometimes see dust dancing in the air. Explain why the dust moves.

1.2 Solids, liquids, and gases

It is easy to tell the difference between a solid, a liquid and a gas:

A solid has a definite shape and a definite volume.

A liquid flows easily. It has a definite volume but no definite shape. Its shape depends on the container.

A gas has neither a definite volume nor a definite shape. It completely fills its container. It is much lighter than the same volume of solid or liquid.

Water: solid, liquid and gas

Water can be a solid (ice), a liquid (water) and a gas (water vapour or steam). Its state can be changed by heating or cooling:

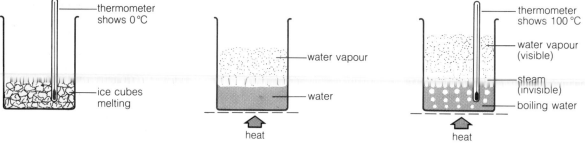

1 **Ice** slowly changes to **water**, when it is put in a warm place. This change is called **melting**. The thermometer shows 0 °C until all the ice has melted, so 0 °C is called its **melting point**.

2 When the water is heated its temperature rises, and some of it changes to **water vapour**. This change is called **evaporation**. The hotter the water gets, the more quickly it evaporates.

3 Soon bubbles appear. The water is **boiling**. Water vapour forms faster. It is now called **steam**. The thermometer shows 100 °C until all the water has changed to steam. 100 °C is the **boiling point** of water.

And when steam is cooled, the opposite changes take place:

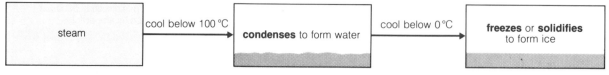

You can see that:
  condensing is the opposite of evaporating;
  freezing is the opposite of melting;
  the freezing point of water is the same as the melting point of ice, 0 °C.

Other things can change state too

Even iron and diamond can melt and boil! Some melting and boiling points are given in this table. Notice how different they can be.

Substance	Melting point/°C	Boiling point/°C
Oxygen	−219	−183
Ethanol	−15	78
Sodium	98	890
Sulphur	119	445
Iron	1540	2900
Diamond	3550	4832

A few substances change straight from solid to gas, without becoming liquid, when they are heated. This change is called **sublimation**. Carbon dioxide and iodine both sublime.

In this play, the 'fog' was carbon dioxide, subliming from lumps on the stage floor.

Melting and boiling points give useful clues

The melting and boiling points of a substance depend on the particles in it. This means that:
1 No two substances have the same melting and boiling points.
2 The melting and boiling point of a substance will change if you mix even a tiny amount of another substance with it.

So you can find out quite a lot from melting and boiling points:
1 You can use them to **identify** a substance.
2 You can use them to say whether it is a **pure** substance or whether it has something else mixed with it. For example:

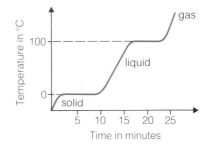

This graph is called a **heating curve**. It shows how the temperature of a substance changes when you heat it. The temperature stays constant while the solid melts and the liquid boils. What is the substance?

This substance melts at 119 °C and boils at 445 °C. Can you identify it? (Look in the table above.)

You need some pure water for an experiment. This water freezes at about −2 °C and boils at about 101 °C. Is it pure?

The water in the beaker is obviously not pure. It has other things mixed with it. These are called **impurities**. And this is what happens:
1 An impurity *lowers* the freezing point of a substance and *raises* its boiling point.
2 The more impurity there is in a substance, the more its freezing and boiling points change.

Questions

1 Write down two properties of a solid, two of a liquid, and two of a gas.
2 What word means the opposite of *boiling*?
3 Which has the lower freezing point, oxygen or ethanol?

4 What useful things can you tell from the melting and boiling points of a substance?
5 One sample of ethanol boils at around 79 °C. Another boils at around 81 °C. How can you tell they are not pure? Which one is *least* pure?

1.3 Particles in solids, liquids, and gases

You saw on page 9 that a substance can change from solid to liquid to gas. The individual *particles* of the substance are the same in each state. It is their *arrangement* that is different.

State	How the particles are arranged	Diagram of particles
Solid	The particles in a solid are packed tightly in a fixed pattern. There are strong forces holding them together, so they cannot leave their positions. The only movements they make are tiny vibrations to and fro.	
Liquid	The particles in a liquid can move about and slide past each other. They are still close together but are not in a fixed pattern. The forces that hold them together are weaker than in a solid.	
Gas	The particles in a gas are far apart, and they move about very quickly. There are almost no forces holding them together. They collide with each other and bounce off in all directions.	

Changes of state

Melting   When a solid is heated, its particles get more energy and vibrate more. This makes the solid **expand**. At the melting point, the particles vibrate so much that they break away from their positions. The solid becomes a liquid.

solid　　heat energy →　　the particles vibrate more　　heat energy at melting point →　　a liquid is formed

Boiling   When a liquid is heated, its particles get more energy and move faster. They bump into each other more often and bounce further apart. This makes the liquid expand. At the boiling point, the particles get enough energy to overcome the forces holding them together. They break away from the liquid and form a gas.

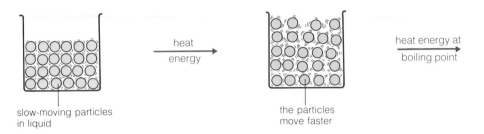

slow-moving particles in liquid

the particles move faster

the particles get enough energy to escape

Evaporating   Some particles in a liquid have more energy than others. Even when a liquid is well below boiling point, *some* particles have enough energy to escape and form a gas. This is called **evaporation**. It is why puddles of rain dry up in the sunshine.

Condensing and solidifying   When a gas is cooled, the particles lose energy. They move more and more slowly. When they bump into each other, they do not have enough energy to bounce away again. They stay close together and a liquid forms. When the liquid is cooled, the particles slow down even more. Eventually they stop moving, except for tiny vibrations, and a solid forms.

Compressing a gas

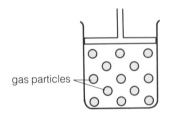

gas particles

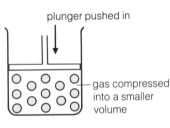

plunger pushed in

gas compressed into a smaller volume

There is a lot of space between the particles in a gas. You can force the particles closer by pushing in the plunger.

The gas gets squeezed or **compressed** into a smaller volume. We say gases are **compressible**.

If enough force is applied to the plunger, the particles get so close together that the gas turns into a liquid. But liquids and solids cannot be compressed because their particles are already close together.

These divers carry compressed air so that they can breathe underwater.

Questions

1 Using the idea of particles, explain why:
   a it is easy to pour a liquid;
   b a gas will completely fill any container;
   c a solid expands when it is heated.

2 Draw a diagram to show what happens to the particles in a liquid, when it boils.
3 Explain why a gas can be compressed into a smaller volume, but a solid can't.

1.4 A closer look at gases

When you blow up a balloon, you fill it with air particles moving at speed. The particles knock against the sides of the balloon and exert **pressure** on it. The pressure is what keeps the balloon inflated. In the same way, *all* gases exert pressure. The pressure depends on the **temperature** of the gas and the **volume** it fills, as you will see below.

When you blow air into a balloon the gas particles exert pressure on the balloon and make it inflate. The more you blow, the greater the pressure.

How gas pressure changes with temperature

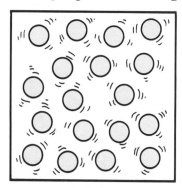 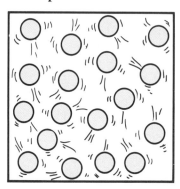

The particles in this gas are moving at speed. They hit the walls of the container and exert pressure on them.
If the gas is heated ...

... the particles take in heat energy and move even faster. They hit the walls more often and with more force. So the gas pressure increases.

The same happens with all gases:
If the volume of a gas is kept constant, its pressure increases with temperature.

How gas pressure changes with volume

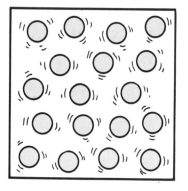 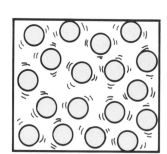

In a pressure cooker, water vapour is heated to well over 100 °C. So it reaches very high pressure. It is dangerous to open a pressure cooker without first letting it cool.

Here again is the gas from above. Its pressure is due to the particles colliding with the walls of the container.

This time the gas is squeezed into a smaller volume. So the particles hit the walls more often. The gas pressure increases.

The same thing is true for all gases:
When a gas is squeezed into a smaller volume, its pressure increases.

How gas volume changes with temperature

Now let's see what happens if the gas pressure is kept constant, but the temperature changes:

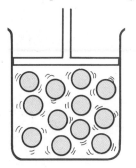 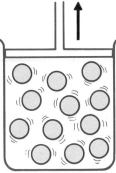

The plunger in this container can move in and out. When the gas is heated, the particles hit it more often . . .

. . . and with more energy, so it moves out. This means the pressure doesn't change. But now the gas fills a larger volume.

This shows that:
If the pressure of a gas is constant, its volume increases with temperature.

The diffusion of gases

On page 7 you saw that gases **diffuse**. A particle of ammonia gas has about half the mass of a particle of hydrogen chloride gas. So will it diffuse faster? Let's see:

1 Cotton wool soaked in ammonia solution is put into one end of a long tube. It gives off ammonia gas.
2 *At the same time*, cotton wool soaked in hydrochloric acid is put into the other end of the tube. It gives off hydrogen chloride gas.
3 The gases diffuse along the tube. White smoke forms where they meet.

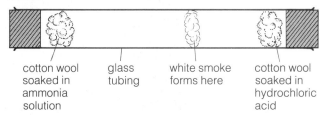

cotton wool soaked in ammonia solution   glass tubing   white smoke forms here   cotton wool soaked in hydrochloric acid

The white smoke forms closer to the right-hand end of the tube. So the ammonia particles have travelled further than the hydrogen chloride particles, in the same length of time.
The lighter the particles of a gas, the faster the gas will diffuse.

The sweet smell of freesias can fill a room. 'Smells' are caused by gas particles diffusing through the air. They dissolve in moisture in the lining of your nose. Then cells in the lining send a message to your brain.

Questions

1 What causes the pressure in a gas?
2 Why does a balloon burst if you keep on blowing?
3 A gas is in a sealed container. How do you think the pressure will change if the container is cooled? Explain your answer.

4 A gas flows from one container into a larger one. What happens to its pressure? Draw diagrams to explain.
5 Of all gases, hydrogen diffuses fastest. What can you tell from that?

1.5 Mixtures (I)

A **mixture** contains more than one substance. So all the substances above are mixtures.

Solutions

A mixture of sugar in water is clear, and you cannot see the sugar. A mixture like this is called a **solution**. We say the sugar has **dissolved**. It is the **solute**, and water is the **solvent**:

solute + solvent = solution

When sugar and water are mixed, the water particles get between the sugar particles and separate them. The separate particles are too small to be seen, which is why the solution looks clear:

A mixture of sugar and water.

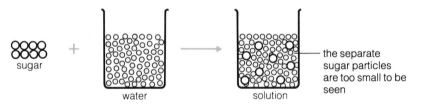

Suspensions

Look at the chalk in water on the right. There are white specks in the water because the chalk has not dissolved — it is **insoluble** in water. A mixture like this is called a **suspension**.
In a suspension, the particles of solid do not all separate. Instead they stay in clusters that are large enough to be seen.

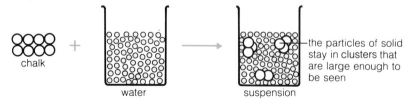

If the clusters of particles are heavy, they sink to the bottom and form a **sediment**.

A mixture of chalk and water.

Other solvents

When water is the solvent, the solution is called an **aqueous solution** (from *aqua*, the Latin word for water).

Water is the most common solvent, but many others are used in industry and about the house. They are needed to dissolve substances that are insoluble in water. Some examples are:

Solvent	It dissolves
White spirit	Gloss paint
Propanone	Grease, nail polish
Dichloromethane	Grease. Used in dry-cleaners
Trichloroethene	Grease and oil. Used in engineering works to 'degrease' metal parts
Ethanol	Glues, printing inks, the scented substances used in perfumes and aftershaves

All the solvents above evaporate easily at room temperature — they are **volatile**. This means that glues and paints dry easily. Aftershaves feel cool because ethanol cools the skin when it evaporates.

Other kinds of solution

Lots of solids dissolve in water to give solutions. So do liquids! For example:

a little washing-up liquid + water = a solution
a little ethanol + water = a solution

But if you add cooking oil to water you won't get a solution. The oil stays as a separate layer. Oil and water don't mix — they are **immiscible**.

Many gases dissolve in water. Carbon dioxide is dissolved in fizzy drinks and bottled drinking water to give them sparkle. The small amount of oxygen dissolved in river water is enough to keep fish alive.

Ethanol is used as a solvent for perfume. It is a volatile liquid. Why is that an advantage?

'Liquid' papers contain a volatile liquid.

Questions

1 Explain each term:
   solution   aqueous solution
   suspension   sediment
2 Name two solids other than chalk that are insoluble in water.
3 At home you make several solutions. Write down:
   a three solids you dissolve in water;
   b three solutions you make by dissolving a liquid in water.
4 What problems might there be if oxygen was *very* soluble in water?
5 Some gases *are* very soluble in water. Try to think of one you have come across in the lab.
6 Name two solvents other than water that are used in the home. What are they used for?
7 Is it a solution or suspension?
   a lemonade   **b** Milk of Magnesia
   c shampoo   **d** typists' correction fluid

1.6 Mixtures (II)

Milk, mayonnaise and hair mousse are all mixtures. But they're not clear, which means they're not solutions. And you can't see the separate bits in them, so they're not suspensions. So what are they?

They all belong to a class of mixtures called **disperse systems**. In a disperse system, insoluble bits of one substance are evenly spread or **dispersed** through another. Let's look at some examples.

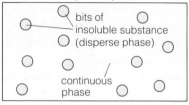

A disperse system

bits of insoluble substance (disperse phase)

continuous phase

Foams

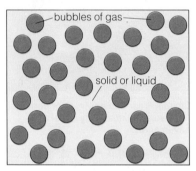

bubbles of gas

solid or liquid

In foams, bubbles of gas (usually air) are dispersed through a solid or liquid.

Hair mousse is a foam. It consists of tiny air bubbles dispersed in a liquid cream.

Expanded polystyrene is a **solid foam** of tiny air bubbles in polystyrene.

Emulsions

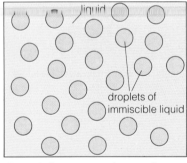

liquid

droplets of immiscible liquid

An emulsion is made up of two immiscible liquids. Tiny drops of one are dispersed in the other.

Eye make-up remover is usually an emulsion of oil droplets in water.

But for greasy, oily hands, you'd use a heavier emulsion of water droplets in oil.

Emulsifiers   Vinegar is a solution of ethanoic acid and other substances in water. When you shake oil and vinegar together to make salad dressing, an emulsion forms. But after a time it separates into two layers again. So the salad dressings you buy in shops usually have an **emulsifying agent** or **emulsifier** added, to stop this happening.

Emulsifiers are made up of long molecules. One end is attracted to oil, the other to water. The molecules surround the oil droplets and keep them dispersed.

a molecule of emulsifier

this end attracted to oil

oil droplet

this end attracted to water

Gels

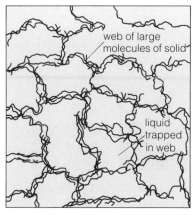

Some solids form a **gel** when mixed with liquid. They have large molecules that attract each other, forming a web. This traps the liquid, making it less runny.

Wallpaper paste is a gel made of cellulose and water.
It thickens when you leave it to stand. It 'dries out' on the wall as the water slowly evaporates.

Jelly is a gel of gelatine and water, with colour and flavour added. Jelly 'sets' as the gelatine web forms. For a wobblier jelly, add more water!

The best of both worlds . . .

Nearly all cosmetics, sauces and household cleaners are disperse systems, and they combine the useful properties of both parts of the system. For example, the tiny air bubbles in hair mousse means it is easier to spread lightly and evenly through your hair, than a liquid would be. And the combination of air and plastic in expanded polystyrene makes it both light and rigid: ideal for packaging.

Most make-up is oil-based. So why can't you remove it with water alone? Why does make-up remover usually contain oil? In what way do you think the water helps?

It's called emulsion paint – but it's not an emulsion! It is a suspension of pigment in a solution of resin in water. The water evaporates, leaving the resin to bind the pigment.

Questions

1 What kind of disperse system is it?
   a foam rubber **b** hair gel
   c meringue **d** mayonnaise
   e milk **f** whipped cream
2 Why is it not possible to make an emulsion from ethanol and water?
3 What is an emulsifier?

4 Washing-up liquid acts as an emulsifier in greasy water. Draw a diagram to show how it works.
5 Explain how the properties of air and rubber make foam rubber suitable for cushions.
6 Wallpaper paste starts off thin and runny, but thickens when it's left to stand. Is this a help or a hindrance when you're putting up wallpaper? Why?

1.7 More about solutions

Helping a solute dissolve

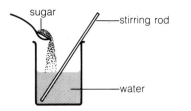

Sugar dissolves quite slowly in water at room temperature (20 °C). Stirring helps!

But if you keep on adding sugar to the water, even with plenty of stirring . . .

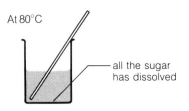

. . . eventually no more will dissolve. The solution is **saturated**.

A saturated solution is one that can dissolve no more solute at that temperature.
But look what happens when you heat the sugar solution:

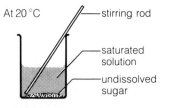

There is undissolved sugar at the bottom of the beaker.

Now some of it has dissolved – but there is still some left.

It has all dissolved. You might even be able to add more!

So sugar is **more soluble** in hot water than in cold. In fact this is usually the case with solids that dissolve. Think about instant coffee for example!
If a solid is soluble in a liquid, it usually gets more soluble as the temperature rises.

Comparing different solutes

The solubility of a substance depends on the particles in it, and how they are held together. So the solubility of every solute is different. For example:

15 grams of this dissolves in 100 grams of water at 50 °C.

39 grams of this dissolves in 100 grams of water at 50 °C.

80 grams of this dissolves in 100 grams of water at 50 °C.

So potassium nitrate is *over five times* as soluble as potassium sulphate, and sodium chloride is nearly *three times* as soluble as it.

Comparing different solvents

The solubility of a solute also depends on the **solvent**.

Iodine is very slightly soluble in water. Only 0.3 grams will dissolve in 100 grams of water at 20 °C. So you get a very pale solution!

It is much more soluble in **cyclohexane**, a solvent which smells like petrol. 2.8 grams of iodine dissolves in 100 grams of cyclohexane at 20 °C.

If you shake some cyclohexane with a solution of iodine in water, almost all the iodine leaves the water and moves into the cyclohexane.

So cyclohexane is much better than water at separating iodine particles from each other! The iodine particles are more attracted to cyclohexane than they are to water.

The solubility of gases

You saw opposite that most *solid* solutes get *more* soluble in water as the temperature rises. The opposite is true for gases. Look at this table.

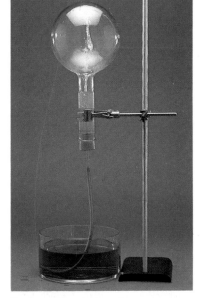

| Gas | Solubility (ml per 100 ml of water) at ... | | | |
	0 °C	20 °C	40 °C	60 °C
Nitrogen	2.4	1.6	1.3	1.0
Oxygen	4.8	3.3	2.5	1.9
Carbon dioxide	171	92.3	56.6	36.0
Sulphur dioxide	7980	4250	2170	
Hydrogen chloride	50 500	47 400	44 500	42 000

So at room temperature (20 °C), hydrogen chloride is over *fourteen thousand* times as soluble as oxygen! It is so soluble that you can do the **fountain experiment** with it.

Nitrogen and oxygen do not react with water when they dissolve in it. But the other three gases react to give acidic solutions.
Sulphur dioxide is a harmful gas given out by car exhausts and power-station chimneys. It dissolves in rainwater to give **acid rain**.

The fountain experiment!

Questions

1 What is a *saturated* solution?
2 A solution of sugar in water is *just* saturated at 20 °C. You heat it to 50 °C. Will it still be saturated? Explain.
3 Why is the solubility of every solute different?
4 Which is more soluble in water at room temperature:
   a oxygen or nitrogen?   **b** oxygen or carbon dioxide?

5 When you heat water, dissolved nitrogen and oxygen bubble off. Will the bubbles contain more nitrogen or more oxygen? Explain.
6 Carbon dioxide is used to make drinks fizzy.
   a In drinks factories, the gas is dissolved in the drinks *under pressure*. Why is this?
   b Why should you keep fizzy drinks in the fridge?

1.8 Separating mixtures (I)

Suppose you need only one substance from a mixture. How do you separate it from the mixture? Here are some methods.

How to separate a solid from a liquid

By filtering   Chalk can be separated from water by filtering the suspension through filter paper. The chalk gets trapped in the filter paper while the water passes through. The chalk is called the **residue**. The water is called the **filtrate**.

Other suspensions can be separated in the same way.

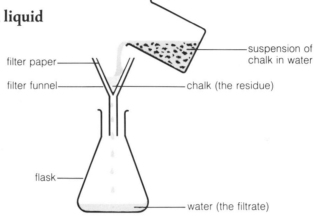

filter paper

filter funnel

flask

suspension of chalk in water

chalk (the residue)

water (the filtrate)

By centrifuging   A centrifuge is used to separate *small* amounts of suspension. In a centrifuge, test-tubes of suspension are spun round very fast, so that the solid gets flung to the bottom:

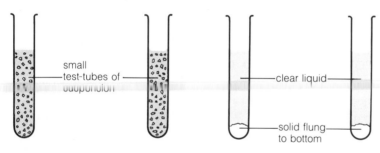

small test-tubes of suspension

clear liquid

solid flung to bottom

A centrifuge.

Before centrifuging, the solid is mixed all through the liquid.

After centrifuging, all the solid has collected at the bottom.

The liquid can be **decanted** (poured out) from the test-tubes, or removed with a small pipette. The solid is left behind.

By evaporating the solvent   If the mixture is a *solution*, the solid cannot be separated by filtering or centrifuging. This is because it is spread all through the solvent in tiny particles. Instead, the solution is heated so that the solvent evaporates, leaving the solid behind. Salt is obtained from its solution by this method:

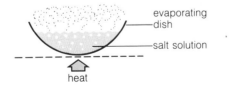

evaporating dish

salt solution

heat

the water evaporates leaving the salt behind

Evaporating the water from a salt solution.

By crystallizing   You can separate many solids from solution by letting them form crystals. Copper(II) sulphate is an example:

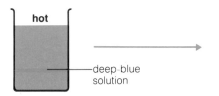

This is a saturated solution of copper(II) sulphate in water at 70 °C. If it is cooled to 20 °C...

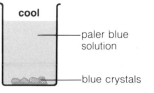

... crystals begin to appear, because the compound is *less soluble* at 20 °C than at 70 °C.

The process is called **crystallization**. It is carried out like this:

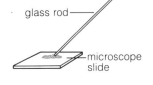

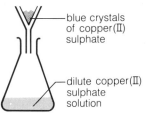

1 A solution of copper(II) sulphate is heated, to get rid of some water. As the water evaporates, the solution becomes more concentrated.

2 The solution can be checked to see if it is ready by placing one drop on a microscope slide. Crystals should form quickly on the cool glass.

3 Then the solution is left to cool and crystallize. The crystals are removed by filtering, rinsed with water and dried with filter paper.

How to separate a mixture of two solids

By dissolving one of them   A mixture of salt and sand can be separated like this:

1 Water is added to the mixture, and it is stirred. The salt dissolves.
2 The mixture is then filtered. The sand is trapped in the filter paper, but the salt solution passes through.
3 The sand is rinsed with water and dried in an oven.
4 The salt solution is evaporated to dryness.

This method works because salt is soluble in water but sand is not. Water could *not* be used to separate salt and sugar, because it dissolves both. You could use ethanol, because it dissolves sugar but not salt. Ethanol is inflammable, so it should be evaporated from the sugar solution over a water bath, as shown on the right.

Evaporating ethanol from a sugar solution, over a water bath.

Questions

1 What is: a filtrate?   a residue?
2 Describe two ways of separating the solid from the liquid in a suspension.
3 Sugar *cannot* be separated from sugar solution by filtering. Explain why.

4 What happens when a solution is evaporated?
5 Describe how you would crystallize potassium nitrate from its aqueous solution.
6 How would you separate salt and sugar? Mention any special safety precaution you would take.

1.9 Separating mixtures (II)

How to separate the solvent from a solution

By simple distillation This is a way of getting pure solvent out of a solution. The apparatus is shown on the right. It could be used to obtain pure water from salt water, for example. This is what happens:

1 The solution is heated in the flask. It boils, and steam rises into the condenser. The salt is left behind.
2 The condenser is cold, so the steam condenses to water in it.
3 The water drips into the beaker. It is completely pure. It is called **distilled water**.

This method could also be used to obtain pure water from sea water, or from ink.

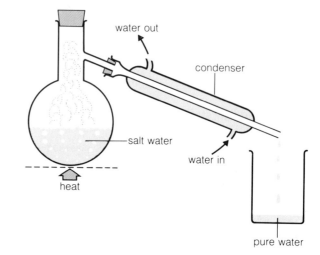

How to separate two liquids

Using a separate funnel If two liquids are **immiscible**, they can be separated with a separating funnel. For example, when a mixture of oil and water is poured into the funnel, the oil floats to the top, as shown on the right. When the tap is opened, the water runs out. The tap is closed again when all the water has gone.

By fractional distillation If two liquids are miscible, they must be separated by fractional distillation. The apparatus is shown below. It could be used to separate a mixture of ethanol and water, for example. These are the steps:

1 The mixture is heated. At about 78 °C, the ethanol begins to boil. Some water evaporates too, so a mixture of ethanol vapour and water vapour rises up the column.
2 The vapours condense on the glass beads in the column, making them hot.
3 When the beads reach about 78 °C, ethanol vapour no longer condenses on them. Only the water vapour does. The water drips back into the flask, while the ethanol vapour is forced into the condenser.
4 There it condenses. Liquid ethanol drips into the beaker.
5 Eventually, the thermometer reading rises above 78 °C. This is a sign that all the ethanol has been separated, so heating can be stopped.

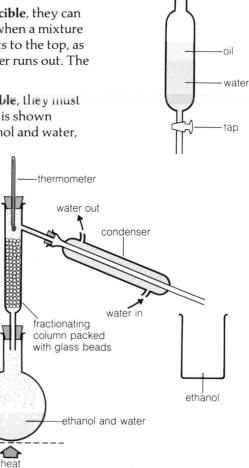

How to separate a mixture of coloured substances

Paper chromatography This method can be used to separate a mixture of coloured substances. For example, it will separate the coloured substances in black ink:

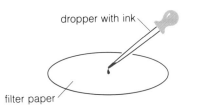

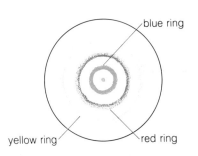

1 A small drop of black ink is placed at the centre of a piece of filter paper, and allowed to dry. Three or four more drops are added on the same spot.

2 Water is then dripped on to the ink spot, one drop at a time. The ink slowly spreads out and separates into rings of different colours.

3 Suppose there are three rings: yellow, red and blue. This shows that the ink contains three substances, coloured yellow, red and blue.

The filter paper with its coloured rings is called a **chromatogram**. In the example above, the outside ring is yellow. This shows that the yellow substance is the most soluble in water. When the water drips on to the ink spot, it dissolves the yellow substance the most easily, and carries it furthest away. Can you tell which substance is the least soluble?

Paper chromatography is often used to **identify** the substances in a mixture. For example, mixture X is thought to contain the substances A, B, C, and D, which are all soluble in propanone. The mixture could be checked like this:

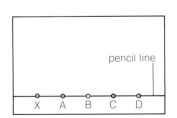

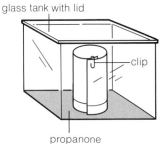

1 Concentrated solutions of X, A, B, C, and D are made up in propanone. A spot of each is placed on a line, on a sheet of filter paper, and labelled.

2 The paper is stood in a little propanone, in a covered glass tank. The solvent rises up the paper; when it's near the top, the paper is taken out again.

3 X has separated into three spots. Two are at the same height as A and B, so X must contain substances A and B. Does it also contain C and D?

Questions

1 How would you obtain pure water from ink? Draw the apparatus you would use, and explain how the method works.
2 Why are condensers so called? What is the reason for running cold water through them?
3 Water and turpentine are immiscible. How would you separate a mixture of the two?
4 Explain how fractional distillation works.
5 In the chromatogram above, how can you tell that X does not contain substance C?

Questions on Section 1

1 A large crystal of potassium manganate(VII) was placed in the bottom of a beaker of cold water and left for several hours.

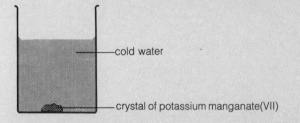

-cold water

-crystal of potassium manganate(VII)

   a Describe what would be seen after five minutes.
   b Describe what would be seen after several hours.
   c Explain your answers using the idea of particles.
   d Name the *two* processes which have taken place during the experiment.

2 Describe the arrangement of particles in a solid, a liquid and a gas.

3 Draw diagrams to show what happens to the particles when:
   a water freezes to ice;
   b steam condenses to water.

4 The graph below is a **heating curve** for a pure substance. It shows how the temperature rises with time, when the solid is heated until it melts, and then the liquid is heated until it boils.

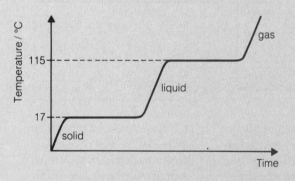

   a What is the melting point of the substance?
   b What is its boiling point?
   c What happens to the temperature while the substance changes state?
   d How can you tell that the substance is not water?

5 Sketch the heating curve for pure water, between − 10 °C and 110 °C. Mark in the temperatures at which water changes state, and its state for each sloping part of the graph.

6 A **cooling curve** shows how the temperature of a substance changes with time, as it is cooled from a gas to a solid. Below is the cooling curve for one substance:

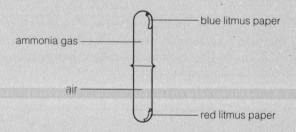

   a What is the state of the substance at room temperature (20 °C)?
   b Use the list of melting and boiling points on page 9 to identify the substance.

7 A test-tube of ammonia gas is placed above a test-tube of air. Ammonia is an alkaline gas that turns litmus blue. It is lighter than air.

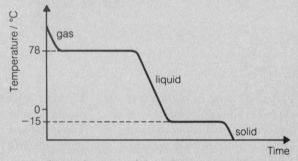

ammonia gas

blue litmus paper

air

red litmus paper

   a After a short time the red litmus paper turns blue. Explain why.
   b Would it make any difference if you reversed the test-tubes? Explain your answer.
   c What would you *see* if the test-tube of air was replaced by one containing hydrogen chloride?

8 In a café, sugar is provided on the tables as sugar cubes. Each cube contains 5 grams of sugar. The café manager notices that there is sugar left in the bottom of many of the used coffee cups. She asks you to carry out an experiment to see how long it takes 5 grams of sugar to dissolve.
   a Make a list of all the variables that affect the time taken for sugar to dissolve in water.
   b In your experiment, what will you keep the same all the way through? Why?
   c Design an experiment to investigate the effect of the variables on your list. Describe it in detail.
   d What results would you expect to obtain?
   e What advice would you give the café manager for solving the problem of the wasted sugar?

9 In dialysis, a membrane is used to separate dispersed particles. A membrane is like filter paper, in that they both act as barriers with holes in them.

In experiment A below, milk passes through the filter paper unchanged. In B, the water remains clear even after several hours.

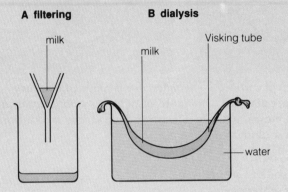

a Name the membrane used in B.
b From the results above, what can you deduce about the holes in the filter paper compared with the holes in the membrane?

After an hour, the water in the beaker was tested and found to contain sugar.
c i Where did the sugar come from?
  ii What can you infer about the size of the sugar particles?
d Milk contains fat, sugar and water particles. Draw a particle diagram to explain to a friend how the dialysis experiment works. Your diagram should show the membrane under high magnification, and all the different particles in the milk.

10 Milk is an emulsion. In milk, droplets of fat are evenly dispersed through water. The fat particles are large enough to be seen under a microscope.
a What does *dispersed* mean?
b Make a drawing of what you would expect to see under the microscope.
c Would you expect the fat particles to be moving or still? Explain your answer.
d Why is milk described as an *emulsion* rather than a *suspension*?

11 At 20 °C (room temperature) 75 ml of ammonia will dissolve in 100 ml of water.
a Look at the table on page 19. How does the solubility of ammonia compare with that of:
  i oxygen?   **ii** hydrogen chloride?
b What can you say about the solubility of ammonia at: **i** 0 °C?   **ii** 40 °C?
c Explain how the fountain experiment works.
d Would you expect the fountain experiment to work with ammonia? Why?
e Would you expect ammonia to have the same solubility in other solvents? Why?

12 Describe the relationship that exists between:
a gas volume and pressure
b gas volume and temperature
c gas pressure and temperature

13

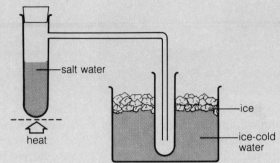

The apparatus above can be used to obtain pure water from salt water.
a What is the purpose of the ice-cold water?
b Why must the glass arm from the first tube reach far down into the second tube?
c Explain how the method works.
d What is this separation method called?

14 A mixture of salt and sugar has to be separated, using the solvent ethanol.
a Which of the two substances is soluble in ethanol?
b Draw a diagram to show how you would separate the salt.
c How could you obtain sugar crystals from the sugar solution, *without* losing the ethanol in the process?
d Draw a diagram of the apparatus for **c**.

15 Eight coloured substances were spotted on to a piece of filter paper, which was then stood in a covered glass tank containing a little propanone. Three of the substances were the basic colours red, blue and yellow. The others were dyes, labelled A, B, C, D, E. The resulting chromatogram is below:

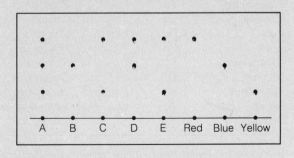

a Which dye contains only one basic colour?
b Which dye contains all three basic colours?
c Which basic colour is most soluble in propanone?

2.1 Atoms, elements, and compounds

Atoms

The pieces of sodium in this photograph are made of billions of tiny particles called **sodium atoms**.

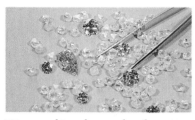

Diamond is a form of carbon. These diamonds are made of billions of **carbon atoms**, which are different from sodium atoms.

Mercury is made of **mercury atoms**, which are different from both sodium atoms and carbon atoms.

Single atoms are far too small to be seen, even with the most powerful microscope. For example, about four billion sodium atoms would fit side by side on the full stop at the end of this sentence. However, in spite of the small size of atoms, scientists have managed to find out a great deal about them. They have found that every atom consists of a **nucleus**, and a cloud of particles called **electrons** that whizz non-stop round the nucleus.

The drawing on the right shows what a sodium atom might look like, greatly magnified.

(You can find out more about the nucleus and electrons on page 28.)

●————————nucleus

————————electron cloud

Elements

Sodium is made of sodium atoms only, so it is an **element**.
An element is a substance that is made of only one kind of atom.
Diamond (carbon) and mercury are also elements.

Altogether, 105 different elements are known. Of these, 90 have been obtained from the Earth's crust and atmosphere, and 15 have been artificially made by scientists.

Every element has a name and a symbol. Here are some of them:

Element	Symbol	Element	Symbol
Aluminium	Al	Bromine	Br
Copper	Cu	Carbon	C
Iron	Fe	Chlorine	Cl
Lead	Pb	Hydrogen	H
Magnesium	Mg	Nitrogen	N
Mercury	Hg	Oxygen	O
Potassium	K	Phosphorus	P
Silver	Ag	Sulphur	S
Sodium	Na	Silicon	Si

A few elements, such as copper, have been known for thousands of years. But most were discovered in the last 400 years. This painting shows Henry Brand, the German alchemist, on his discovery of phosphorus in 1659. He extracted it from urine, by accident, during his search for the elxir of life. To his amazement it glowed in the dark.

It is easy to remember that the symbol for <u>a</u>luminium is Al, and for <u>c</u>arbon is C. But some symbols are harder to remember, because they are taken from the Latin names for the elements. For example: potassium has the symbol K, from its Latin name <u>k</u>alium. Sodium has the symbol Na, from its Latin name <u>na</u>trium.

The metals The elements in the list on page 26 are in two columns for a good reason. The ones on the left are **metals**, while those on the right are **nonmetals**. Over 80 of the elements are metals. Although they all look different, they have many properties in common. Here are a few:

1 They allow electricity and heat to pass through them easily – they are all good **conductors** of electricity and heat.
2 They are all solids at room temperature, except mercury, and most of them have high melting points.
3 Most of them can be hammered into different shapes (they are **malleable**) and drawn into wires (they are **ductile**).

The nonmetals Only about one-fifth of the elements are nonmetals. They are quite different from metals:

1 They are poor conductors of electricity and heat. (Carbon is an exception.)
2 They usually have low melting points (eleven of them are gases and one is a liquid at room temperature).
3 When solid nonmetals are hammered, they break up – they are **brittle**.

Nerves of steel? 16 elements (including iron, but mainly oxygen, carbon, nitrogen, calcium and phosphorus) combine to make the hundreds of compounds in the human body.

Compounds

Elements can combine with each other to form **compounds**. **A compound contains atoms of different elements joined together.** Although there are only 105 elements, there are millions of compounds. This table shows three common ones:

Name of compound	Elements in it	How the atoms are joined up
Water	Hydrogen and oxygen	O / H H
Carbon dioxide	Carbon and oxygen	O C O
Ethanol	Carbon, hydrogen and oxygen	H H / H C C O H / H H

Symbols for compounds The symbol for a compound is called its **formula**. It is made up from the symbols of the elements. The formula for water is H_2O and for ethanol is C_2H_5OH. Can you see why? Can you guess the formula for carbon dioxide? (Check your answer on page 78.) Note that the plural of **formula** is **formulae**.

Questions

1 What is an atom?
2 What is the centre part of an atom called?
3 Explain what an element is.
4 Explain what these words mean:
   malleable   ductile   brittle

5 Write down three properties of nonmetals.
6 What is: **a** a compound?   **b** a formula?
7 What is H_2O? What does the $_2$ in it show?
8 A certain compound has the formula NaOH. What elements does it contain?

2.2 More about the atom

Protons, neutrons and electrons

On page 26 you saw that all atoms consist of a **nucleus** and a cloud of **electrons** that move round the nucleus. The nucleus is itself a cluster of two sorts of particle, **protons** and **neutrons**.
All the particles in an atom are very light. Their mass is measured in **atomic mass units**, rather than grams. Protons and electrons also have an **electric charge**:

Particle in atom	Mass	Charge
Proton ●	1 unit	Positive charge ($+1$)
Neutron ●	1 unit	None
Electron ●	Almost nothing	Negative charge (-1)

How the particles are arranged

The sodium atom is a good one to start with. It has **11** protons, **11** electrons and **12** neutrons. They are arranged like this:

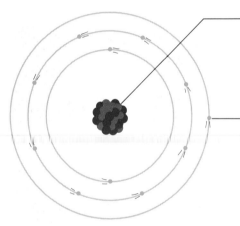

the protons and neutrons cluster together in the centre, forming the nucleus; this is the heavy part of the atom

the electrons circle very fast around the nucleus, at different levels from it.

The nucleus is very tiny compared with the rest of the atom. If the atom was the size of a football stadium, the nucleus (sitting on the centre spot) would be the size of a pea!

The different energy levels for the electrons are called **electron shells**. Each shell can hold only a limited number of electrons:

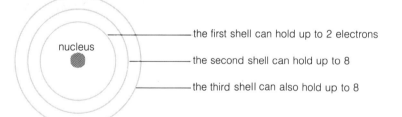

nucleus

the first shell can hold up to 2 electrons

the second shell can hold up to 8

the third shell can also hold up to 8

Notice how the electrons are arranged in the sodium atom:
  2 in the first shell (it is full);
  8 in the second shell (it is also full);
  1 in the third shell (it is not full).
The atom is often written as **Na(2,8,1)**. Can you see why?
The (2,8,1) is its electron arrangement or **electronic configuration**.

Niels Bohr, a Danish scientist, was the first person to put forward the idea of electron shells. He died in 1962.

Proton number and mass number

Proton number   Look again at the sodium atom on the opposite page. It has **11** protons. This fact could be used to identify it, because *only* a sodium atom has 11 protons. Every other atom has a different number of protons.
You can identify an atom by the number of protons in it.
The number of protons in an atom is called its **proton number** or **atomic number**.
The proton number of sodium is 11.

The sodium atom also has **11** electrons. So it has an equal number of protons and electrons. The same is true for every sort of atom:
Every atom has an equal number of protons and electrons.
Because of this, atoms have no overall charge. The charge on the electrons cancels the charge on the protons. You can check this for a sodium atom, on the right.

Mass number   The electrons in an atom have almost no mass. So the mass of an atom is nearly all due to its protons and neutrons. For this reason, the number of protons and neutrons in an atom is called its **mass number** or **nucleon number**.
The mass number = the number of protons + neutrons in an atom.
A sodium atom has 11 protons and 12 neutrons, so the mass number of sodium is 23.

Since the proton number is the number of protons only, then:
     mass number − proton number = number of neutrons.
So, for a sodium atom, the number of neutrons = (23−11) = 12.

The charge on a sodium atom:
●●●● 11 protons
●●●● Each has a charge of +1
●●● Total charge +11
●●●● 11 electrons
●●●● Each has a charge of −1
●●● Total charge −11
Adding the charges: +11
−11
0
The answer is zero. The atom has no overall charge.

Shorthand for an atom

The sodium atom can be described in a short way, using:
     the symbol for sodium (Na)
     its proton number (11)
     its mass number (23)

The information is written as $^{23}_{11}\text{Na}$.

From it you can tell that the sodium atom has 11 protons, 11 electrons and 12 neutrons (23−11=12). Chemists often describe atoms in this way. The information is always put in the same order:

     mass number
     proton number   **symbol**

A hundred years ago, hardly anything was known about the atom. For example, the neutron was discovered only in 1932, by this British scientist, Sir James Chadwick.

Questions

1 Name the particles that make up the atom.
2 Which particle has:
   a a positive charge?   **b** no charge?
   c almost no mass?
3 Draw a sketch of the sodium atom.
4 What does *electronic configuration* mean?

5 What does the term mean?   **a** proton number
   b mass number   **c** nucleon number
6 Name each of these atoms, and say how many protons, electrons and neutrons it has:

   $^{12}_{6}\text{C}$ $^{16}_{8}\text{O}$ $^{24}_{12}\text{Mg}$ $^{27}_{13}\text{Al}$ $^{64}_{29}\text{Cu}$

2.3 Some different atoms

What makes two atoms different?

On page 28 you saw that sodium atoms have 11 protons. This is what makes them different from all other atoms. *Only* sodium atoms have 11 protons, and any atom with 11 protons *must* be a sodium atom.

In the same way, an atom with 6 protons must be a carbon atom, and an atom with 7 protons must be a nitrogen atom:

You can identify an atom by the number of protons in it.

The first twenty elements There are 105 elements altogether. Of these, hydrogen has the smallest atoms, with only 1 proton each. Helium atoms have 2 protons each, lithium atoms have 3 protons each, and so on up to hahnium atoms, which have 105 protons each. Below are the first twenty elements, arranged in order according to the number of protons they have:

Element	Symbol	Number of protons (proton number)	Number of electrons	Number of neutrons	Number of protons + neutrons (mass number)
Hydrogen	H	1	1	0	1
Helium	He	2	2	2	4
Lithium	Li	3	3	4	7
Beryllium	Be	4	4	5	9
Boron	B	5	5	6	11
Carbon	C	6	6	6	12
Nitrogen	N	7	7	7	14
Oxygen	O	8	8	8	16
Fluorine	F	9	9	10	19
Neon	Ne	10	10	10	20
Sodium	Na	11	11	12	23
Magnesium	Mg	12	12	12	24
Aluminium	Al	13	13	14	27
Silicon	Si	14	14	14	28
Phosphorus	P	15	15	16	31
Sulphur	S	16	16	16	32
Chlorine	Cl	17	17	18	35
Argon	Ar	18	18	22	40
Potassium	K	19	19	20	39
Calcium	Ca	20	20	20	40

Drawing the different atoms

It is easy to draw the different atoms, if you remember these rules:
1 The protons and neutrons form the nucleus at the centre.
2 The electrons are in electron shells around the nucleus.
3 The first electron shell can hold a maximum of 2 electrons, the second can hold 8, and the third can also hold 8.
In the drawings below, **p** = proton, **e** = electron and **n** = neutron.

A hydrogen atom 1p, 1e	A lithium atom 3p, 3e, 4n	A magnesium atom 12p, 12e, 12n
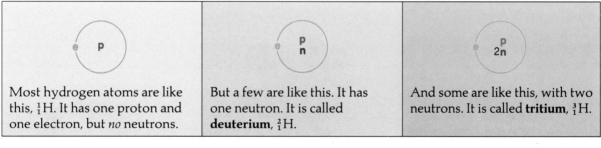		
Note that it has no neutrons. So its proton number is 1 *and* its mass number is 1. It is described as $^1_1$H.	It has three electrons. The first shell can hold only two electrons, so a second shell is used. The atom is described as $^7_3$Li. Can you explain why?	The first two shells together hold only ten electrons. The remaining two electrons must go in a third shell. The atom is described as $^{24}_{12}$Mg. Can you explain why?

Isotopes

The atoms of an element are not always identical:

Most hydrogen atoms are like this, $^1_1$H. It has one proton and one electron, but *no* neutrons.	But a few are like this. It has one neutron. It is called **deuterium**, $^2_1$H.	And some are like this, with two neutrons. It is called **tritium**, $^3_1$H.

All three atoms belong to the element hydrogen, *because they all have 1 proton*. They are called **isotopes** of hydrogen.
Isotopes are atoms of the same element, with the same number of protons but different numbers of neutrons.

Most elements have more than one isotope. For example carbon has three: $^{12}_6$C, $^{13}_6$C and $^{14}_6$C.   Chlorine has two: $^{35}_{17}$Cl and $^{37}_{17}$Cl.

Questions

1 Which element has atoms with:
   1 5 protons?   **b** 15 protons?   **c** 20 protons?
2 A hydrogen atom can be described as $^1_1$H. Use the same method to describe the atoms of all the elements listed on page 30.

3 Draw a sketch of: **a** a lithium atom;
   b a carbon atom;   **c** a neon atom.
   Give the electronic configuration for each.
4 What is an isotope? Name the isotopes of hydrogen and write symbols for them.

2.4 The periodic table

Arranging the elements in groups

Two groups of elements are shown below. For each element, the electronic configuration of its atoms is given.
The three elements in each group have something in common:

Group 1 The atoms of these elements each have 1 electron in the outer shell.	**Group 2** The atoms of these elements each have 2 electrons in the outer shell.
Lithium Li (2,1)	Beryllium Be (2,2)
Sodium Na (2,8,1)	Magnesium Mg (2,8,2)
Potassium K (2,8,8,1)	Calcium Ca (2,8,8,2)

Using the same idea, scientists have put all the elements into groups.

1 First, the elements are listed in order of increasing proton number. Hydrogen comes first, since its proton number is 1.
2 Next, the list is sub-divided like this:
   The elements whose atoms have 1 outer-shell electron are picked out of the list, in order, and called **Group 1**.
   The elements whose atoms have 2 outer-shell electrons are picked out and called **Group 2**.
   Group 3, **Group 4** and so on are formed in the same way.
3 Then the groups are arranged side by side to give the **periodic table**. This is shown opposite and again on page 250.

The Russian chemist Mendeleev, who drew up the first version of the periodic table in 1869.

The periodic table

Group

	1	2													3	4	5	6	7	0
1					$^{1}_{1}$H hydrogen															$^{4}_{2}$He helium
2	$^{7}_{3}$Li lithium	$^{9}_{4}$Be beryllium													$^{11}_{5}$B boron	$^{12}_{6}$C carbon	$^{14}_{7}$N nitrogen	$^{16}_{8}$O oxygen	$^{19}_{9}$F fluorine	$^{20}_{10}$Ne neon
3	$^{23}_{11}$Na sodium	$^{24}_{12}$Mg magnesium					The transition metals								$^{27}_{13}$Al aluminium	$^{28}_{14}$Si silicon	$^{31}_{15}$P phosphorus	$^{32}_{16}$S sulphur	$^{35\cdot5}_{17}$Cl chlorine	$^{40}_{18}$Ar argon
4	$^{39}_{19}$K potassium	$^{40}_{20}$Ca calcium	$^{45}_{21}$Sc scandium	$^{48}_{22}$Ti titanium	$^{51}_{23}$V vanadium	$^{52}_{24}$Cr chromium	$^{55}_{25}$Mn manganese	$^{56}_{26}$Fe iron	$^{59}_{27}$Co cobalt	$^{59}_{28}$Ni nickel	$^{64}_{29}$Cu copper	$^{65}_{30}$Zn zinc	$^{70}_{31}$Ga gallium	$^{73}_{32}$Ge germanium	$^{75}_{33}$As arsenic	$^{79}_{34}$Se selenium	$^{80}_{35}$Br bromine	$^{84}_{36}$Kr krypton
5	$^{85}_{37}$Rb rubidium	$^{88}_{38}$Sr strontium	$^{89}_{39}$Y yttrium	$^{91}_{40}$Zr zirconium	$^{93}_{41}$Nb niobium	$^{96}_{42}$Mo molybdenum	$^{98}_{43}$Tc technetium	$^{101}_{44}$Ru ruthenium	$^{103}_{45}$Rh rhodium	$^{106}_{46}$Pd palladium	$^{108}_{47}$Ag silver	$^{112}_{48}$Cd cadmium	$^{115}_{49}$In indium	$^{119}_{50}$Sn tin	$^{122}_{51}$Sb antimony	$^{128}_{52}$Te tellurium	$^{127}_{53}$I iodine	$^{131}_{54}$Xe xenon
6	$^{133}_{55}$Cs caesium	$^{137}_{56}$Ba barium	$^{139}_{57}$La lanthanum	$^{178\cdot5}_{72}$Hf hafnium	$^{181}_{73}$Ta tantalum	$^{184}_{74}$W tungsten	$^{186}_{75}$Re rhenium	$^{190}_{76}$Os osmium	$^{192}_{77}$Ir iridium	$^{195}_{78}$Pt platinum	$^{197}_{79}$Au gold	$^{201}_{80}$Hg mercury	$^{204}_{81}$Tl thallium	$^{207}_{82}$Pb lead	$^{209}_{83}$Bi bismuth	$^{210}_{84}$Po polonium	$^{210}_{85}$At astatine	$^{222}_{86}$Rn radon
7	$^{223}_{87}$Fr francium	$^{226}_{88}$Ra radium	$^{227}_{89}$Ac actinium															

$^{140}_{58}$Ce cerium	$^{141}_{59}$Pr prae-sodymium	$^{144}_{60}$Nd neodymium	$^{147}_{61}$Pm promethium	$^{150}_{62}$Sm samarium	$^{152}_{63}$Eu europium	$^{157}_{64}$Gd gadolinium	$^{159}_{65}$Tb terbium	$^{162}_{66}$Dy dysprosium	$^{165}_{67}$Ho holmium	$^{167}_{68}$Er erbium	$^{169}_{69}$Tm thulium	$^{173}_{70}$Yb ytterbium	$^{175}_{71}$Lu lutecium
$^{232}_{90}$Th thorium	$^{231}_{91}$Pa prot-actinium	$^{238}_{92}$U uranium	$^{237}_{93}$Np neptunium	$^{242}_{94}$Pu plutonium	$^{243}_{95}$Am americium	$^{247}_{96}$Cm curium	$^{247}_{97}$Bk berkelium	$^{251}_{98}$Cf californium	$^{254}_{99}$Es einsteinium	$^{253}_{100}$Fm fermium	$^{256}_{101}$Md mendelevium	$^{254}_{102}$No nobelium	$^{257}_{103}$Lw lawrencium

The groups The table has eight groups of elements, plus a block of **transition metals**. The eight groups are numbered. Group 4 contains the elements carbon (C), silicon (Si), germanium (Ge), tin (Sn), and lead (Pb). Their atoms each have 4 electrons in the outer shell. The atoms of Group 5 elements each have 5 electrons in the outer shell, and so on. Now look at the last group, Group 0. Their atoms all have *full outer shells*.

Some of the groups have special names:

    Group 1 is often called **the alkali metals.**
    Group 2 is **the alkaline earth metals.**
    Group 7 is **the halogens.**
    Group 0 is **the noble gases**.

Look at the zigzag line through the groups. It separates the **metals** from the **nonmetals**. The metals are on the left.

The periods The horizontal rows in the table are called **periods**. Period 2 contains lithium (Li), beryllium (Be), boron (B), carbon (C), nitrogen (N), oxygen (O), fluorine (F), and neon (Ne).

The transition metals The atoms of these have more complicated electron arrangements. Note that the group contains many common metals, such as iron (Fe), nickel (Ni), and copper (Cu).

One of the elements in the periodic table is named after this famous scientist. Can you work out which one?

Questions

1 Explain why beryllium, magnesium and calcium are all in Group 2 of the periodic table.

2 Copy and complete:
  a In Group 4, the atoms have 4 . . .
  b In Group . . ., the atoms have 6 . . .
  c In Group 0, the atoms have . . .

3 What are the rows in the table called?

4 Use the larger table on page 250 to help you name the elements in: **a** Group 5 **b** Period 1 **c** Period 3

5 What is the special name for the elements in:
  a Group 1? **b** Group 7? **c** Group 0?

6 Draw a large outline of the periodic table and mark in the names and symbols for the first twenty elements. Use the table on page 250 to help you.

2.5 A closer look at the periodic table (I)

On the last two pages you saw that the periodic table shows all the elements arranged in groups.

A group of elements is sometimes called a **family**, because its elements resemble each other. Sometimes they look alike, and usually they behave alike. As you'll see, their behaviour depends mainly on the number of electrons in their atoms' outer shells.

Group 0 – the noble gases

This group contains the elements helium, neon, argon, krypton and xenon. These elements are all:

- nonmetals;
- gases (they occur naturally in the air you breathe);
- colourless.

The striking thing about them is how unreactive they are. They will not normally react with anything. (In fact, until 1962 it was thought they were completely unreactive. Then scientists discovered that some could be forced to react weakly with other substances.)

Why they have similar properties These elements have similar properties because their atoms have full outer electron shells. **A full outer electron shell makes an atom unreactive.**
Atoms that don't have a full outer shell react with other atoms in order to obtain one. **The noble gases are unreactive because their atoms have full outer electron shells.**

Looking for trends The noble gases are not *identical*, however. For example if you filled five balloons with the same volume of each gas at the same temperature and let them go, this is what you'd find:

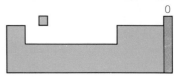

All from the same family. Can you see the resemblance? Like humans, the elements in a family or **group** resemble each other.

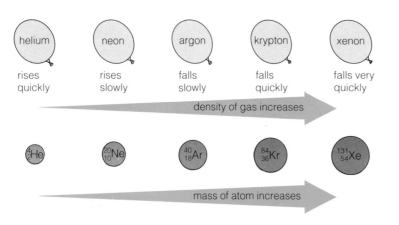

The balloon experiment shows that the 'heaviness' or **density** of the gases increases from helium to xenon. This is an example of a gradual change or **trend** in the periodic table.
As you go down Group 0, the density of the gases increases.

The density increases because the mass of the atoms increases. The atoms also get larger, but that change is not so great.

Helium is lighter than air.

Group 7 – the halogens

Chlorine, bromine and iodine are the most common halogens. They are all nonmetals, and poisonous. They all have coloured vapours: chlorine is a green gas, bromine forms a red vapour and iodine a purple one. They also react in similar ways. For example with iron:

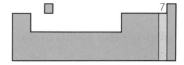

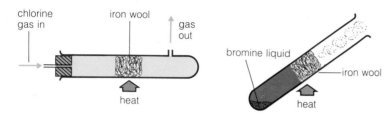

Hot iron wool glows brightly when chlorine passes over it. Brown smoke forms, and a brown solid is left behind.

The iron glows less brightly when bromine is used. Brown smoke and a brown solid are formed.

With iodine, the iron glows even less brightly. But once again, brown smoke and a brown solid are formed.

Chlorine is the most reactive of these three halogens – it reacts the most easily with iron. Iodine is the least reactive of the three.

Why they have similar properties The halogens have similar properties because their atoms all have 7 electrons in the outer shell.

Why they are reactive The halogens are reactive because their atoms are just one electron short of a full outer shell. They can gain this electron by reacting with atoms of other elements.

Trends within the group Look at these trends:

Atom	Element		Melts at ... (°C)	Boils at ... (°C)	
$^{35\cdot5}_{17}Cl$	chlorine (green gas)	reactivity decreases as atoms get larger	−101	−35	melting and boiling points increase
$^{80}_{35}Br$	bromine (red liquid)		−7	59	
$^{127}_{53}I$	iodine (black solid)		114	184	

All the halogens exist as molecules. Melting and boiling points *increase* as we go down the group because the attraction between molecules increases. More energy is needed to help them escape from the solid to form liquid, and from the liquid to form gas.

At the same time, reactivity *decreases* because of size. A halogen atom is able to attract an extra electron into its outer shell because of the positive charge on the nucleus. (Opposite charges attract.) But as the atoms get bigger, their outer shells get further from the nucleus. The force of attraction gets less. So the element gets less reactive.

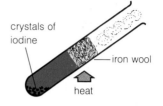

chlorine gas in   heat

iron wool

iron wool   crystals of iodine

heat   heat

gas out   iron wool

bromine liquid

Questions

1 a Why is a group of elements often called a family?
  b Why do its members have similar properties?
2 Explain why: a the noble gases are unreactive;
  b their density increases as we go down Group 0.
3 Explain why the halogens are reactive elements.

4 The first element in Group 7 is fluorine. Would you expect it to be:
  a a gas, a liquid or a solid? Why?
  b coloured or colourless?   c harmful or harmless?
  d more reactive or less reactive than chlorine?

2.6 A closer look at the periodic table (II)

Group 1 – the alkali metals

The first three elements of Group 1 are lithium, sodium, and potassium. All three:
- are soft metals that can be cut with a knife.
- are so light they float on water.
- are silvery and shiny when freshly cut, but quickly tarnish.
- have low melting and boiling points compared with other metals.

They also react in a similar way. For example:

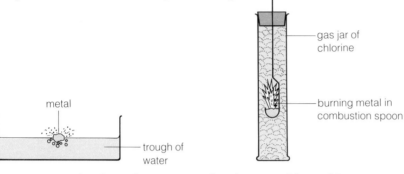

They react violently with water to give hydrogen and an alkaline solution. Lithium floats, and sodium shoots across the water, while potassium melts and the hydrogen catches fire.

They burn quickly in chlorine with a bright flame. Potassium burns fastest and sodium next. White solids are left behind. These solids are potassium, sodium and lithium chloride.

Lithium is the least reactive of the three metals – it reacts the most slowly. Potassium is the most reactive of the three.

Why they have similar properties   Alkali metals have similar properties because their atoms all have 1 electron in the outer shell.

Trends within the group   Look at these trends:

Atom	Metal		Melts at . . . (°C)	Boils at . . . (°C)	
$^{7}_{3}$Li	lithium		181	1342	
$^{23}_{11}$Na	sodium	reactivity increases as atoms get larger	98	883	melting and boiling points decrease
$^{39}_{19}$K	potassium		63	760	
$^{85}_{37}$Rb	rubidium		39	686	

As we go down Group 1, reactivity increases while melting and boiling points decrease. (This is the opposite of Group 7.)
Alkali metals react to *lose* an outer electron and obtain a full outer shell. The further the electron is from the nucleus, the easier this is. So the bigger the atom, the more reactive the metal will be. Meanwhile melting and boiling points *decrease* because the attraction between the atoms gets less strong as the atoms get larger.

Group 2 – the alkaline earth metals

The elements in Group 2 are also silvery metals. But if you compare them with Group 1 metals in the same period, you'll find differences:

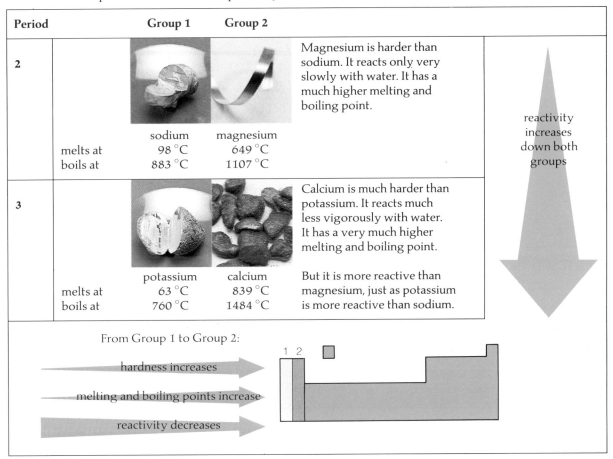

Period		Group 1	Group 2		
2		sodium	magnesium	Magnesium is harder than sodium. It reacts only very slowly with water. It has a much higher melting and boiling point.	reactivity increases down both groups
	melts at	98 °C	649 °C		
	boils at	883 °C	1107 °C		
3		potassium	calcium	Calcium is much harder than potassium. It reacts much less vigorously with water. It has a very much higher melting and boiling point. But it is more reactive than magnesium, just as potassium is more reactive than sodium.	
	melts at	63 °C	839 °C		
	boils at	760 °C	1484 °C		

From Group 1 to Group 2:

hardness increases

melting and boiling points increase

reactivity decreases

Why alkaline earth metals are less reactive   When Group 2 metals react they have to give up *two* outer electrons to obtain a full outer shell. This is more difficult than losing just one electron, so they are less reactive than Group 1 metals.

But as the atoms get bigger it gets easier to lose the two electrons, so the metals get more reactive as you go down the group. This was also the trend in Group 1.

Now look at the melting and boiling points for Group 2, shown on the right. Some of the values seem out of place. But the *overall* trend is a decrease as you go down the group.

Groups 1 and 2 share two trends: increasing reactivity down the group, and an overall decrease in melting and boiling points.

Group 2	Melting point in °C	Boiling point in °C
Beryllium	1278	2970
Magnesium	649	1107
Calcium	839	1484
Strontium	769	1384
Barium	725	1640

The trend is not so smooth. (Magnesium seems out of place, for example.) But the overall trend is a decrease from top to bottom.

Questions

1 Sodium is more reactive than lithium. Why?
2 Caesium comes below rubidium in Group 1. Compared with rubidium, would you expect it:
   a to have a higher or a lower melting point?
   b to be more reactive or less reactive?

3 Why do the Group 2 metals have similar properties?
4 Magnesium is less reactive than sodium. Why?
5 Lithium is the first element in Group 1 and beryllium the first in Group 2. Which do you think:
   a is more reactive?   **b** has a higher melting point?

2.7 A closer look at the periodic table (III)

Trends across a period

Look at the elements from Period 3 of the periodic table:

Group	1	2	3	4	5	6	7	0
Element	sodium	magnesium	aluminium	silicon	phosphorus	sulphur	chlorine	argon
Outer electrons	1	2	3	4	5	6	7	8
Element is a ...	metal	metal	metal	metalloid	nonmetal	nonmetal	nonmetal	nonmetal
Reactivity	high ————————————→ low ————————————→ high							unreactive
Melting point (°C)	98	649	660	1410	590	119	−101	−189
Boiling point (°C)	883	1107	2467	2355	(ignites)	445	−35	−186
Oxide formed ...	Na_2O	MgO	Al_2O_3	SiO_2	P_2O_5	SO_3	Cl_2O_7	none
Oxide is ...	basic	amphoteric					acidic	–

There are several trends to notice as you move across the period:

1 The number of outer-shell electrons increases by 1 each time. (It is the same as the group number.)
2 The elements go from metals to nonmetals. Silicon is in between, like a metal in some ways and a nonmetal in others. It is called a **metalloid**.
3 The melting and boiling points increase to the middle of the period, then decrease again. They are lowest on the right. (Only chlorine and argon are gases at room temperature.)
4 All the elements except argon react with oxygen to form oxides. (Why doesn't argon?) The ratio of oxygen atoms to metal atoms in the oxide increases steadily across the period.
5 The oxides on the left are **basic**, which means they react with acids to form salts. Those on the right are **acidic** – they react with alkalis to form salts. Aluminium oxide is in between – it reacts with both acids and alkalis to form salts. It is called an **amphoteric oxide**. (There is more about oxides on page 174.)

The elements in Period 2 show similar trends. Here beryllium has an amphoteric oxide, and boron and carbon are metalloids. Carbon behaves as a nonmetal in its reactions. But as **graphite** it is a very good conductor of electricity, like a metal! (The other form of carbon, diamond, does not conduct.)

Period 3 of the periodic table.

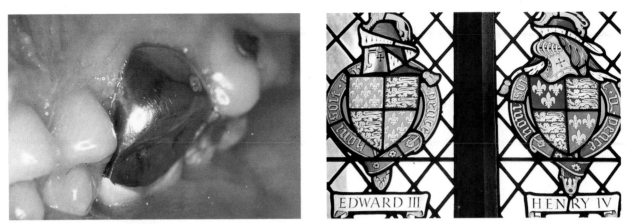

Transition metals provide, among other things, the 'colour' for glass, the lead that holds it together, and gold for teeth.

The transition metals

The long block in the middle of the periodic table contains metals only. These are the **transition metals**. They include most of the hard, dense and less reactive metals used for building things.
Iron, lead, and copper are transition metals. So are gold, platinum and silver. These last three are called **precious metals** because they are rare, beautiful and costly, and so unreactive that they can be used as jewellery.

Transition metals and their compounds are often used in industry as catalysts to speed up reactions. (See page 124.) They are also used to make alloys. Transition metal compounds are usually coloured. (Compounds of Group 1 and 2 metals are white.)

Hydrogen

Hydrogen stands on its own in the periodic table. This is because it has one outer electron, like the Group 1 metals, but unlike them it is a gas, and it usually reacts like a nonmetal.

Some artificial elements

Not all the elements occur naturally. Fifteen are artificial, created by scientists during nuclear reactions. Most of these are in the bottom block of the periodic table, and include the elements neptunium (Np) to lawrencium (Lr) in the very bottom row.

All the isotopes of the artificial elements are **radioactive**. This means their atoms have an unstable nucleus which breaks down, giving out radiation. You can find out more about this on the next few pages.

Questions

1 A challenge! Make a table like the one opposite, but for *Period 2*. Show the group numbers, names of elements and numbers of outer electrons. Now try to predict:
   a the melting and boiling points for the elements;
   b the formulae for their oxides.

2 Name ten transition metals. (Page 33 will help!) Now write down one use for each of the ten.
3 Most paints contain compounds of transition elements. Why do you think this is?
4 One transition metal is a liquid at room temperature. Which one?

2.8 Radioactivity (I)

Atoms with an unstable nucleus

Many atoms behave in a way that has nothing to do with their outer electrons. For example carbon has three isotopes, $^{12}_6C$, $^{13}_6C$ and $^{14}_6C$, called carbon-12, carbon-13 and carbon-14:

 $^{12}_6C$ 6 protons
6 electrons
6 neutrons

 $^{13}_6C$ 6 protons
6 electrons
7 neutrons

 $^{14}_6C$ 6 protons
6 electrons
8 neutrons

An atom of carbon-12.
Almost 99% of carbon atoms
are like this.

An atom of carbon-13.
Just over 1% of carbon atoms
are like this.

An atom of carbon-14.
A tiny percentage of carbon
atoms are like this.

The carbon-14 atom is the strange one. Its nucleus is **unstable** because of the extra neutrons. Sooner or later, every single carbon-14 atom throws a particle out of its nucleus and becomes a *nitrogen* atom. This incredible process is called **decay**.

Carbon-14 is said to be **radioactive**. It is called a **radioisotope** or a **radionuclide**. When it decays it gives out **radiation**.

Many elements have radioactive isotopes. Elements at the bottom of the periodic table often have several, because their atoms have lots of neutrons. For example radon, the last element in Group 0, has two isotopes and both are radioactive. Sooner or later they decay.
All radioisotopes eventually turn into stable atoms by giving out radiation.

For carbon-14, the decay has just one stage. But for some isotopes it has several. For example polonium-216 decays like this:

| polonium-216 | lead-212 | bismuth-212 | thallium-208 | lead-208 |
| radioactive | radioactive | radioactive | radioactive | stable |

Decay is a random process

It could take just a few seconds, or hundreds of years, for a given atom of polonium-216 to turn into an atom of lead-208. That's because decay is random. You can't control it by heat, or catalysts, or chemicals, because it is a *nuclear* reaction, not a chemical one.

Radioisotopes decay even when they are bonded to other atoms. *Every* carbon-14 atom decays to form a nitrogen atom, no matter whether it is in a chunk of charcoal, or in a molecule of carbon dioxide gas, or in a molecule of glucose inside a plant.

Half-life is the time it takes for half the radioisotopes in a sample to decay.

What is radiation?

Radiation consists of three types of particle:

1. **Alpha particles.** An alpha particle consists of 2 protons and 2 neutrons, so it has a mass of 4 units. It shoots out of the radioisotope at high speed but slows down quickly in air. It can't pass through paper or skin.

The radioisotope americium-241 is used in smoke alarms. It has 95 protons, and decays by giving out an alpha particle. That leaves 93 protons, which means it has turned into neptunium:

$$^{241}_{95}\text{Am} \longrightarrow {}^{237}_{93}\text{Np} + \text{an alpha particle}$$

Can you see why the mass number has dropped to 237?

2. **Beta particles.** A beta particle is a fast electron, which means it has almost no mass. It can travel 20 or 30 cm in air, and through skin and thin sheets of metal.
It is formed when a neutron turns into a proton and an electron. The new electron is thrown out. So the *mass* of the isotope does not change but its protons increase by 1.

A carbon-14 atom has 6 protons. It decays by giving off a beta particle. Now it has 7 protons. It has turned into a nitrogen atom:

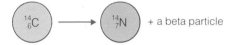

$$^{14}_{6}\text{C} \longrightarrow {}^{14}_{7}\text{N} + \text{a beta particle}$$

3. **Gamma rays.** Gamma rays are usually given out at the same time as alpha and beta rays. They are high-energy rays, like X-rays, and they travel at the speed of light. They can travel several metres in air, and through thick sheets of metal, and deep into your body. But thick sheets of lead or thick concrete will stop them.

A smoke detector containing americium-241. Smoke entering the detector is ionized by the radiation. The ionized smoke conducts an electric current which sets the alarm off.

Radioactive materials are kept under closely controlled conditions.

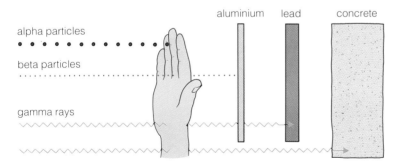

Questions

1. Name the 3 types of radiation and say what they are.
2. Think of a reason why it is easier to slow down alpha particles than beta particles.
3. Explain why the waste material from nuclear power-stations is usually buried deep below ground.

Use the periodic table (page 250) for **4** and **5**.

4. What is formed when an alpha particle is given off by an atom of: **a** uranium-238?
   b plutonium-239? **c** thorium-232?
5. What is formed when a beta particle is given off by an atom of: **a** lead-212?
   b thorium-234? **c** iodine-131?

2.9 Radioactivity (II)

Radiation can be harmful

When radiation collides with molecules in the air, or in blocks of concrete, or in your body, it knocks electrons from them, turning them into charged particles called **ions**. The radiation **ionizes** the molecules. When it happens in body cells, the cells may die. Or they may undergo a change or **mutation**.

A large dose of radiation over a short period of time will kill millions of cells in your gut, and blood cells, and bone tissues. The result is **radiation sickness**: vomiting, tiredness, loss of appetite, hair loss, bleeding gums, and usually death within weeks.

But with small doses over a long period of time, the mutated cells get the chance to multiply. Years later they may show up as **cancer**. Or if your sex cells mutate and you then have children, they may be born with deformities.

Background radiation

Your body can tolerate a *low* level of radiation without ill effects, because it is able to repair any damage caused. This is just as well, since we are surrounded by low-level radiation. We call it **background radiation**.

Background radiation comes from the soil, rocks, the air, water, plants, building materials, and food. Some is caused by the decay of radioisotopes such as carbon-14 that occur naturally, and some is due to the cosmic rays that arrive from outer space. Radiation from rocks is highest in areas where the rock is granite.

Making use of radiation

Although radioactive substances are dangerous and must be handled carefully, they are also very useful. Here are some examples . . .

Tracers Engineers can check underground oil and gas pipes for leaks by adding a radioactive substance to the oil or gas, so that leaks can be detected using a Geiger counter. Radioisotopes used in this way are called **tracers**.

Krypton-81 is used as a tracer in lung checks. You breathe in a little and it decays in your lungs. The radiation shows up as bright spots on a TV screen, and dark patches show where the lungs are not working properly.

Tracers are also used to study how animals digest food, and plants take up fertilizers. For example, a plant is put into soil containing a little phosphorus-32. Then its roots, stem, and leaves are checked at intervals, to see where the phosphorus has got to.

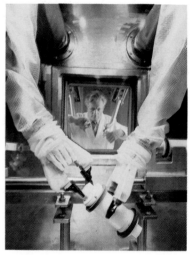

Some elements in the periodic table are dangerous to handle, because they are radioactive. Special safety equipment has to be used.

A Geiger counter being used to check for radioactivity.

Fertilizer take-up can be 'traced' using radioactive phosphorus.

Killing microbes Gamma rays kill the microbes that cause fruit and vegetables to decay. So food irradiated by gamma rays stays fresh on supermarket shelves for longer. In the same way, companies making syringes and surgical instruments use gamma rays to sterilize them after packing, to remove any risk of infection.

Cancer treatment Gamma rays can pass deep into the body and damage body cells, which could eventually lead to cancer. But gamma rays can also be used to kill off cancer cells and so *cure* cancer. This is called **radiotherapy**. The secret is to control the beam of rays so that it is the right strength, and hits its target exactly. Cobalt-60 is often used as the source of gamma rays for radiotherapy.

Carbon dating You saw earlier that carbon-14 atoms decay to form nitrogen atoms. At the same time, high in the upper atmosphere, nitrogen atoms are bombarded by neutrons and turned into carbon-14 atoms. So the amount of carbon-14 around us remains constant.

All living things contain carbon. Plants take it in from carbon dioxide in the air, then animals eat the plants. The carbon always includes the same tiny proportion of carbon-14.

When a plant or animal dies it stops taking in carbon. But its carbon-14 atoms continue to decay. By comparing the radiation from them with that from living things, and knowing the half-life of carbon-14, the age of the remains can be worked out.

This process is called **carbon dating**. It can be used to find the age of anything that was once living material: old pieces of wood or charcoal, plants or shells preserved in rock, ancient manuscripts, and bones.

Dating rocks Potassium has one naturally-occurring radioisotope, called potassium-40. Out of every ten thousand potassium atoms, only twelve are potassium-40. This radioisotope has a half-life of twelve thousand million years! It decays to form a stable argon atom.

Many rocks contain potassium compounds. When the potassium-40 in these compounds decays, the argon that forms stays trapped in the rock. By measuring the potassium and argon in the rock and comparing the two, scientists can work out how old the rock is.

Some people object to the idea of irradiating food to preserve freshness.

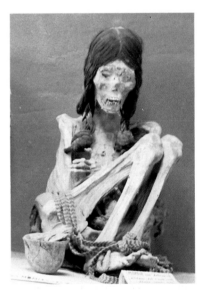
Carbon dating showed that this mummy from Chile was about 1100 years old.

Questions

1 What is *background radiation*?
2 People who fly a lot at high altitudes are exposed to more background radiation than the rest of us. Suggest a reason for this.
3 What is a radioactive *tracer*? Give an example.
4 Phosphorus-30 has a half-life of 2.5 minutes and phosphorus-32 of 14.3 days. Phosphorus-32 is the one chosen as a tracer in plant experiments. Can you explain why?
5 Explain how radioactivity is used to date rock.

Questions on Section 2

1 Turn to page 30 and learn the names and symbols for the first twenty elements, in order. Then close the book and write them out.

2 **a** Give one difference between an element and a compound.
b The formulae of some compounds are given below. Write down the names of the elements they contain.
CO_2 $CaCl_2$ H_2S $PbCO_3$ KOH HgO

3 Hydrogen, deuterium and tritium are isotopes. Their structures are shown below.

hydrogen   deuterium   tritium

a Copy and complete the following key:
● represents . . . .
○ represents . . . .
⊗ represents . . . .
b What are the mass numbers of hydrogen, deuterium and tritium?
c Copy and complete this statement.
Isotopes of an element always contain the same number of . . . . and . . . . but different numbers of . . . .
d The average mass number of naturally-occurring hydrogen is 1.008. Which isotope is present in the highest proportion, in naturally-occurring hydrogen?

4 Copy and complete the following table for isotopes of some common elements:

Isotope	Name of element	Proton number	Mass number	Number of p e n
$^{16}_{8}O$	oxygen	8	16	8 8 8
$^{18}_{8}O$				
$^{12}_{6}C$				
$^{13}_{6}C$				
$^{25}_{12}Mg$				
$^{26}_{12}Mg$				

5 For each of the six elements aluminium (Al), boron (B), nitrogen (N), oxygen (O), phosphorus (P), sulphur (S), write down:
a the period of the periodic table to which it belongs;
b its group number in the periodic table;
c its proton number (atomic number);
d the number of electrons in one atom;
e its electronic configuration;
f the number of outer-shell electrons in one atom.
Which of the above elements would you expect to have similar properties? Why?

6 The statements below are about metals and nonmetals. Say whether each is true or false. (If false, give a reason.)
a All metals conduct electricity.
b All metals are solid at room temperature.
c Nonmetals are good conductors of heat but poor conductors of electricity.
d Many nonmetals are gases at room temperature.
e Most metals are brittle and break when hammered.
f Most nonmetals are ductile.
g There are about four times as many metals as nonmetals.

7 **a** Make a larger copy of this outline of the periodic table:

b Write in the group and period numbers.
c Draw a zigzag line to show how the metals are separated from the nonmetals in the table.
d Now put the letters A to L in the correct places in the table, to fit these descriptions:
A   The lightest element.
B   Any noble gas.
C   The element with proton number 5.
D   The element with 6 electrons in its atoms.
E   Any element with 6 outer-shell electrons.
F   The most reactive alkali metal.
G   The least reactive alkali metal.
H   The most reactive halogen.
I   A Group 3 metal.
J   An alkaline earth metal.
K   A transition metal.
L   A Group 5 nonmetal.

8 This table gives data for some elements:

Name	Symbol	Melting point (°C)	Boiling point (°C)	Electrical conductivity
Aluminium	Al	660	2450	good
Bromine	Br	−7	58	poor
Calcium	Ca	850	1490	good
Chlorine	Cl	−101	−35	poor
Copper	Cu	1083	2600	good
Helium	He	−270	−269	poor
Iron	Fe	1540	2900	good
Lead	Pb	327	1750	good
Magnesium	Mg	650	1110	good
Mercury	Hg	−39	357	good
Nitrogen	N	−210	−196	poor
Oxygen	O	−219	−183	poor
Phosphorus	P	44	280	poor
Potassium	K	64	760	good
Sodium	Na	98	890	good
Sulphur	S	119	445	poor
Tin	Sn	230	2600	good
Zinc	Zn	419	906	good

Use the table to answer these questions.
a What is the melting point of iron?
b Which element melts at −7 °C?
c Which element boils at 280 °C?
d Which element has a boiling point only 1 °C higher than its melting point?
e Over what temperature range is sulphur a liquid?
f Which element has:
  i the highest melting point?
  ii the lowest melting point?
g Divide the elements into two groups, one of metals and the other of nonmetals.
h List the *metals* in order of increasing melting point.
i Which metal is a liquid at room temperature (20 °C)?
j Which nonmetal is a liquid at room temperature?
k List the elements which are gases at room temperature. What do you notice about the elements in this list?

9 Before the development of the periodic table, a scientist called Döbreiner discovered sets of three elements with similar properties, which he called **triads**. Some of these triads are shown as members of the same group, in the modern periodic table. Complete the following triads by inserting the missing middle element:
  chlorine (Cl), ..., iodine (I);
  lithium (Li), ..., potassium (K)
  calcium (Ca), ..., barium (Ba).

10 Many scientists contributed to the development of the modern periodic table. One of them was the Russian chemist Mendeleev. In 1869 he arranged the elements that were then known, in a table very similar to the one in use today. He realized that gaps should be left for elements that had not yet been discovered, and even went so far as to predict the properties of several of these elements.
Rubidium is an alkali metal that lies below potassium in Group 1. Here is some data for Group 1:

Element	Proton number	Melting point (°C)	Boiling point (°C)	Chemical reactivity
Lithium	3	180	1330	quite reactive
Sodium	11	98	890	reactive
Potassium	19	64	760	very reactive
Rubidium	37	?	?	?
Caesium	55	29	690	violently reactive

a Using your knowledge of the periodic table, predict the missing data for rubidium.
b In a rubidium atom:
  i how many electron shells are there?
  ii how many electrons are there?
  iii how many outer-shell electrons are there?

11 This question is about elements from the families called: alkali metals, alkaline earth metals, transition metals, halogens, noble gases.
Element A is a soft, silvery metal which reacts violently in water.
Element B is a gas at room temperature. It reacts violently with other elements, without heating.
Element C is a gas that sinks in air. It does not react readily with any other element.
Element D is a hard solid at room temperature and forms coloured compounds.
Element E conducts electricity and reacts slowly with water. During the reaction its atoms each give up two electrons.
a Place the elements in their correct families. Give further information about the position of the element within the family.
b Describe the outer shell of electrons for each element described above.
c How does the arrangement of electrons in their atoms make some elements very reactive and others unreactive?
d Name elements which fit descriptions A to E.

3.1 Why compounds are formed

Most elements form compounds

On page 36 you saw that sodium reacts with chlorine:

When sodium is heated and placed in a jar of chlorine, it burns with a bright flame.

The result is a white solid, which has to be scraped from the sides of the jar.

The white solid is called **sodium chloride**. It is formed by atoms of sodium and chlorine joining together, so it is a **compound**. The reaction can be described like this:

sodium + chlorine ⟶ sodium chloride

The + means *reacts with*, and the ⟶ means *to form*. Most elements react to form compounds. For example:

lithium + chlorine ⟶ lithium chloride
hydrogen + chlorine ⟶ hydrogen chloride

The noble gases do not usually form compounds

The noble gases are different from other elements, in that they do not usually form compounds, as you saw on page 34. For this reason, their atoms are described as **unreactive** or **stable**. They are stable because their outer electron shells are *full*:
A full outer shell makes an atom stable.

Helium atom, full outer shell: *stable*

Neon atom, full outer shell: *stable*

Argon atom, full outer shell: *stable*

Only the noble gas atoms have full outer shells. The atoms of all other elements have incomplete outer shells. That is why they react. **By reacting with each other, atoms can obtain full outer shells and so become stable.**

Several of the noble gases are used in lighting. For example, xenon is used in lighthouse lamps, like this one. It gives a beautiful blue light.

Losing or gaining electrons

The atoms of some elements can obtain full shells by *losing* or *gaining* electrons, when they react with other atoms:

Losing electrons   The sodium atom is a good example. It has just 1 electron in its outer shell. It can obtain a full outer shell by losing this electron to another atom. The result is a **sodium ion**:

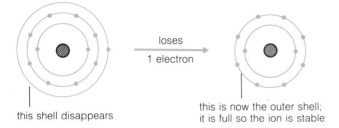

sodium atom (11p, 11e)   **sodium ion (11p, 10e)**

loses
1 electron

this shell disappears

this is now the outer shell;
it is full so the ion is stable

The charge on the sodium ion:	
the charge on 11 protons is	$+11$
the charge on 10 electrons is	-10
total charge	$+1$

The sodium ion has 11 protons but only 10 electrons, so it has a charge of $+1$, as you can see from the box on the right above.

The symbol for sodium is Na, so the sodium ion is called **Na$^+$**. The $^+$ means *1 positive charge*. Na$^+$ is a **positive ion**.

Gaining electrons   A chlorine atom has 7 electrons in its outer shell. It can reach a full shell by accepting just 1 electron from another atom. It becomes a **chloride ion**:

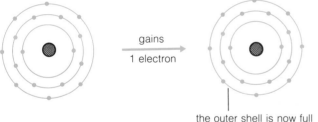

chlorine atom (17p, 17e)   **chloride ion (17p, 18e)**

gains
1 electron

the outer shell is now full
so the ion is stable

The charge on the chloride ion:	
the charge on 17 protons is	$+17$
the charge on 18 electrons is	-18
total charge	-1

The chloride ion has a charge of -1, so it is a **negative ion**. Its symbol is **Cl$^-$**.

Ions   Any atom becomes an ion if it loses or gains electrons. **An ion is a charged particle. It is charged because it contains an unequal number of protons and electrons.**

Questions

1 What is another word for *unreactive*?
2 Why are the noble gas atoms unreactive?
3 Explain why all other atoms are reactive.
4 Draw a diagram to show how a sodium atom obtains a full outer shell.
5 Explain why the sodium ion has a charge of $+1$.

6 Draw a diagram to show how a chlorine atom can obtain a full outer shell.
7 Write down the symbol for a chloride ion. Why does this ion have a charge of -1?
8 Explain what an ion is, in your own words.
9 Why do noble gas atoms *not* form ions?

3.2 The ionic bond

The reaction between sodium and chlorine

As you saw on the last page, a sodium atom can lose one electron, and a chlorine atom can gain one, to obtain full outer shells. So when a sodium atom and a chlorine atom react together, the sodium atom loses its electron *to the chlorine atom*, and two ions are formed. Here, sodium electrons are shown as • and chlorine electrons as ×, but remember that all electrons are exactly the same:

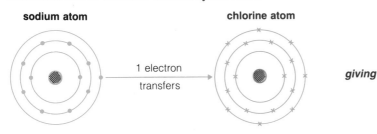

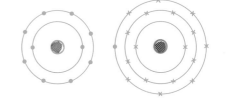

stable ions with full shells

The two ions have opposite charges, so they attract each other. The force of attraction between them is strong. It is called an **ionic bond**, or sometimes an **electrovalent bond**.

How solid sodium chloride is formed

When sodium reacts with chlorine, billions of sodium and chloride ions form, and are attracted to each other. But the ions do not stay in pairs.

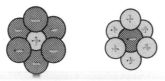

Instead, they cluster together, so that each ion is surrounded by six ions of opposite charge. They are held together by strong ionic bonds. (Each ion forms six bonds.)
The pattern grows until a giant structure of ions is formed. It contains equal numbers of sodium and chloride ions. This giant structure is the compound **sodium chloride**, or **salt**.

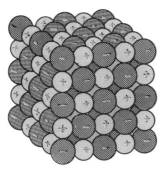

Because sodium chloride is made of ions, it is called an **ionic compound**. It contains one Na^+ ion for each Cl^- ion, so its formula is **NaCl**.
The charges in the structure add up to zero:

the charge on each sodium ion is	$+1$
the charge on each chloride ion is	-1
total charge	0

The compound therefore has no overall charge.

These polystyrene spheres have been given opposite charges, so they are attracted to each other. The same happens with ions of opposite charge.

Other ionic compounds

Sodium is a **metal**, and chlorine is a **nonmetal**. They react together to form an **ionic compound**. Other metals can also react with non-metals to form ionic compounds. Below are two more examples.

Magnesium and oxygen A magnesium atom has 2 outer electrons and an oxygen atom has 6. Magnesium burns fiercely in oxygen. During the reaction, each magnesium atom loses its 2 outer electrons to an oxygen atom. Magnesium ions and oxide ions are formed:

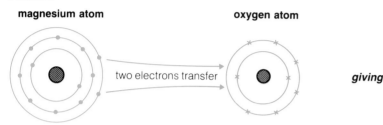

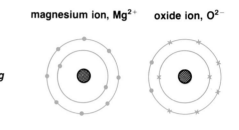

The ions attract each other because of their opposite charges. Like the ions on the last page, they group together into a giant ionic structure. The resulting compound is called **magnesium oxide**. Magnesium oxide contains one magnesium ion for each oxide ion, so its formula is **MgO**. The compound has no overall charge:

the charge on each magnesium ion is   2 +
the charge on each oxide ion is   2 −
                      total charge   0

Magnesium and chlorine To obtain full outer shells, a magnesium atom must lose 2 electrons, and a chlorine atom must gain 1 electron. So when magnesium burns in chlorine, each magnesium atom reacts with *two* chlorine atoms, to form **magnesium chloride**:

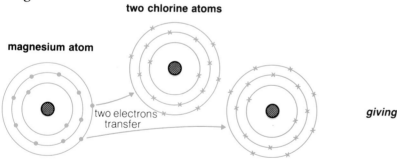

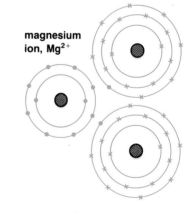

The ions form a giant ionic structure, with *two* chloride ions for each magnesium ion. The formula of magnesium chloride is therefore **MgCl$_2$**. The compound has no overall charge. Can you explain why?

Questions

1 Draw a diagram to show what happens when a sodium atom reacts with a chlorine atom.
2 What is an ionic bond? What is the other name for it?
3 Sketch the structure of sodium chloride, and explain why its formula is NaCl.

4 Explain why:
   a a magnesium ion has a charge of 2 + ;
   b the ions in magnesium oxide stay together;
   c magnesium chloride has no overall charge;
   d the formula of magnesium chloride is MgCl$_2$.

3.3 Some other ions

Ions of the first twenty elements

Not every element forms ions during reactions. In fact, out of the first twenty elements in the periodic table, only eleven easily form ions. These ions are given below, with their names.

Group		H$^+$ hydrogen						0
1	2		3	4	5	6	7	none
Li$^+$ lithium	Be^{2+} beryllium		none	none	none	O^{2-} oxide	F$^-$ fluoride	none
Na$^+$ sodium	Mg^{2+} magnesium		Al^{3+} aluminium	none	none	S^{2-} sulphide	Cl$^-$ chloride	none
K$^+$ potassium	Ca^{2+} calcium	transition metals						

Note that hydrogen and the metals form **positive ions**, which have the same names as the atoms. The nonmetals form **negative ions** and their names end in -*ide*.

The elements in Groups 4 and 5 do not usually form ions, because their atoms would have to gain or lose several electrons, and that takes too much energy. The elements in Group 0 do not form ions because their atoms already have full shells.

The names and formulae of their compounds

The names   To name an ionic compound, you just put the names of the ions together, with the positive one first:

Ions in compound	Name of compound
K$^+$ and F$^-$	Potassium fluoride
Ca^{2+} and S^{2-}	Calcium sulphide

The formulae   The formulae of ionic compounds can be worked out by the following steps. Two examples are given below.

1 Write down the name of the ionic compound.
2 Write down the symbols for its ions.
3 The compound must have no overall charge, so **balance** the ions, until the positive and negative charges add up to zero.
4 Write down the formula without the charges.

Bath salts contain Na$^+$ and CO$_3^{2-}$ ions; Epsom salts contain Mg^{2+} and SO$_4^{2-}$ ions. Can you give the names and formulae of the three main compounds illustrated here?

Example 1
1 Lithium fluoride.
2 The ions are Li$^+$ and F$^-$.
3 One Li$^+$ is needed for every F$^-$, to make the total charge zero.
4 The formula is LiF.

Example 2
1 Sodium sulphide.
2 The ions are Na$^+$ and S^{2-}.
3 Two Na$^+$ ions are needed for every S^{2-} ion, to make the total charge zero: Na$^+$ Na$^+$ S^{2-}.
4 The formula is Na$_2$S. (What does the $_2$ show?)

Transition metal ions

Some transition metals form only one type of ion:
- silver forms only Ag^+ ions
- zinc forms only Zn^{2+} ions

but most of them can form more than one type. For example, copper and iron can each form two:

Ion	Name	Example of compound
Cu^+	copper(I) ion	copper(I) oxide, Cu_2O
Cu^{2+}	copper(II) ion	copper(II) oxide, CuO
Fe^{2+}	iron(II) ion	iron(II) chloride, $FeCl_2$
Fe^{3+}	iron(III) ion	iron(III) chloride, $FeCl_3$

The (II) in a name shows that the ion has a charge of $2+$. What do the (I) and (III) show?

Compound ions

So far, all the ions have been formed from single atoms. But ions can also be formed from groups of joined atoms. These are called **compound ions**. The most common ones are shown on the right. Remember, each is just one ion, even though it contains more than one atom. The formulae for their compounds can be worked out as before. Some examples are shown below.

NH_4^+, the ammonium ion

OH^-, the hydroxide ion

NO_3^-, the nitrate ion

SO_4^{2-}, the sulphate ion

CO_3^{2-}, the carbonate ion

HCO_3^-, the hydrogen carbonate ion

Example 3
1 Sodium carbonate.
2 The ions are Na^+ and CO_3^{2-}.
3 Two Na^+ are needed to balance the charge on one CO_3^{2-}.
4 The formula is Na_2CO_3.

Example 4
1 Calcium nitrate.
2 The ions are Ca^{2+} and NO_3^-.
3 Two NO_3^- are needed to balance the charge on one Ca^{2+}.
4 The formula is $Ca(NO_3)_2$. Note that a bracket is put round the NO_3, before the $_2$ is put in.

Questions

1 Explain why a calcium ion has a charge of $2+$.
2 Why is the charge on an aluminium ion $3+$?
3 Write down the symbols for the ions in:
  a potassium chloride   **b** calcium sulphide
  c lithium sulphide   **d** magnesium fluoride
4 Now work out the formula for each compound in question 3.
5 Work out the formula for each compound:
  a copper(II) chloride   **b** iron(III) oxide
6 Write a name for each compound: $CuCl$, FeS, Na_2SO_4, $Mg(NO_3)_2$, NH_4NO_3, $Ca(HCO_3)_2$
7 Work out the formula for:
  a sodium sulphate   **b** potassium hydroxide
  c silver nitrate   **d** ammonium nitrate

3.4 The covalent bond

Sharing electrons

When two nonmetal atoms react together, *both of them need to gain electrons* to reach full shells. They can manage this only by sharing electrons between them. Atoms can share only their outer electrons, so just the outer electrons are shown in the diagrams below.

Hydrogen A hydrogen atom has only one electron. Its shell can hold two electrons, so is not full. When two hydrogen atoms get close enough, their shells overlap and then they can share electrons:

Because the atoms share electrons, there is a strong force of attraction between them, holding them together. This force is called a **covalent bond**. The bonded atoms form a **molecule**.
A molecule is a small group of atoms which are held together by covalent bonds.
Hydrogen gas is made up of hydrogen molecules, and for this reason it is called a **molecular** substance. Its formula is H_2.
Several other nonmetals are also molecular. For example:

chlorine, Cl_2	iodine, I_2	oxygen, O_2
nitrogen, N_2	sulphur, S_8	phosphorus, P_4

Chlorine A chlorine atom needs a share in one more electron, to obtain a full shell. So two chlorine atoms bond covalently like this:

Oxygen The formula for oxygen is O_2, so each molecule must contain two atoms. Each oxygen atom has only six outer electrons; it needs a share in two more to reach a full shell:

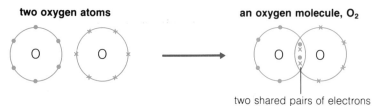

Since the oxygen atoms share two pairs of electrons, the bond between them is called a **double covalent bond**, or just a **double bond**.

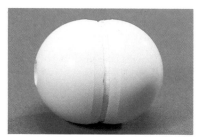

A model of the hydrogen molecule. The molecule can be shown as H—H. The line represents a single bond.

In H_2, the $_2$ tells you the number of hydrogen atoms in each molecule.
Because it has two atoms in each molecule, hydrogen is described as **diatomic**.

A model of the chlorine molecule.

A model of the oxygen molecule. The molecule can be shown as O=O. The lines represent a double bond.

Covalent compounds

On the opposite page you saw that several nonmetal elements exist as molecules. A huge number of compounds also exist as molecules. In a molecular compound, atoms of *different* elements share electrons with each other. These compounds are often called **covalent compounds** because of the covalent bonds in them. Water, ammonia and methane are all covalent compounds.

Water   The formula of water is H_2O. In each molecule, an oxygen atom shares electrons with two hydrogen atoms, and they all reach full shells:

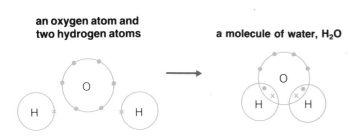

an oxygen atom and
two hydrogen atoms

a molecule of water, H_2O

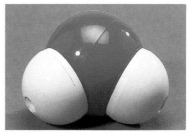

A model of the water molecule.

Ammonia   Its formula is NH_3. Each nitrogen atom shares electrons with three hydrogen atoms, and they all reach full shells:

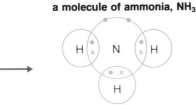

a nitrogen atom and
three hydrogen atoms

a molecule of ammonia, NH_3

A model of the ammonia molecule.

Methane   Its formula is CH_4. Each carbon atom shares electrons with four hydrogen atoms, and they all obtain full shells:

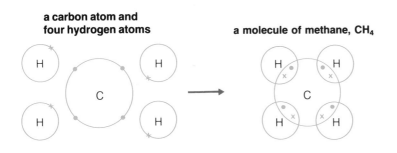

a carbon atom and
four hydrogen atoms

a molecule of methane, CH_4

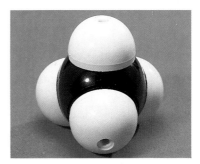

A model of the methane molecule.

Questions

1 What is the name of the bond between atoms that share electrons?
2 What is a molecule?
3 Give five examples of molecular elements.
4 Draw a diagram to show the bonding in:
   a chlorine   **b** oxygen

5 The bond between two oxygen atoms is called a double bond. Why?
6 Show the bonding in a molecule of:
   a water   **b** methane
7 Hydrogen chloride (HCl) is also molecular. Draw a diagram to show the bonding in it.

3.5 Ionic and molecular solids

On page 10 you saw that solids are made of particles packed closely together in a regular pattern. If the particles are **ions**, the solids are called **ionic solids**. If they are **molecules**, the solids are called **molecular** or **covalent** solids. The two types of solid have quite different properties, as you will see below.

Ionic solids

One of the most common ionic solids is sodium chloride – it is ordinary table salt. You already know quite a lot about its structure:

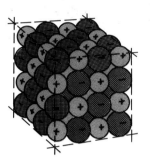

Sodium chloride is made of sodium and chloride ions, packed in a regular pattern. This arrangement is called a **lattice**. The ions are held together by strong **ionic bonds**

The pattern repeats millions of times. The result is a piece of solid with straight edges and flat faces, called a **crystal**. Above is a crystal of sodium chloride, magnified 35 times.

The crystals look white and shiny. A box of table salt contains millions of them.

Sodium chloride is typical of ionic solids. In *all* ionic solids, the ions are packed in a regular pattern, and held together by strong ionic bonds. This means that all ionic solids are **crystalline**. The crystals may be quite large, or so small you need a microscope to see them.

Their properties Ionic solids have these properties:

1 They have high melting and boiling points. For example:
   - sodium chloride melts at 808 °C and boils at 1465 °C.
   - sodium hydroxide melts at 319 °C and boils at 1390 °C.

   This is because ionic bonds are very strong, so it takes a lot of heat energy to break up the lattice and form a liquid. In fact all ionic substances are solid at room temperature.
2 They shatter when hit with a hammer: they are **brittle**.
3 They are usually soluble in water, but insoluble in other solvents such as tetrachloromethane and petrol.
4 They do not conduct electricity when solid. Electricity is a stream of moving charges. Although the ions *are* charged, they cannot move, because the ionic bonds hold them firmly in position. (You will learn more about this in Chapter 7.)
5 They *do* conduct electricity when they are melted or dissolved. This is because the ions are then free to move.

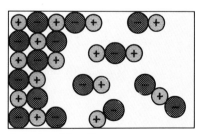

An ionic solid melting

Molecular solids

Iodine is a good example of a molecular solid. Each iodine molecule contains two atoms, held together by a strong covalent bond.

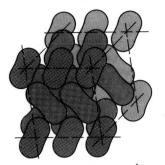

Here ⬮ represents an iodine molecule. The molecules are packed in a regular pattern, and held together by weak forces.

The pattern repeats millions of times, and the result is a crystal. Above is a single iodine crystal, magnified 15 times.

Iodine crystals are grey-black, and shiny, and brittle. This jar contains hundreds of thousands of them.

Iodine is typical of molecular solids. In *all* molecular solids, the molecules are held together in a regular pattern. So the solids are crystalline. The forces that hold the molecules together are weak.

Their properties   Molecular solids have these properties:

1 They have low melting points and boiling points – much lower than ionic solids do. This is because it doesn't take much heat energy to overcome the weak forces between molecules. In fact many molecular substances melt, and even boil, *below* room temperature, so are liquids or gases at room temperature. Here are some examples:

Substance	Melting point/°C	Boiling point/°C
Oxygen	−219	−183
Chlorine	−101	−35
Water	0	100
Naphthalene	80	218

2 They shatter when hit with a hammer: they are brittle.
3 Unlike ionic solids, molecular solids are usually insoluble in water, but soluble in solvents such as tetrachloromethane and petrol.
4 They do not conduct electricity. Molecules are not charged, so molecular substances cannot conduct, even when melted.

Some molecular substances and their state at room temperature:

Solids   iodine
sulphur
naphthalene

Liquids   bromine
water
ethanol

Gases   oxygen
nitrogen
carbon dioxide

Questions

1 What is:
 a an ionic solid?   **b** a molecular solid?
 Give three examples of each.
2 What is another name for molecular solids?
3 Explain how a crystal of sodium chloride is formed. Can you think of a reason why its faces are flat?

4 Why do ionic solids have high melting points?
5 List four properties of covalent solids.
6 Explain why many molecular substances are gases or liquids at room temperature, and give four examples.
7 You can buy solid air-fresheners in shops. Do you think these substances are ionic or covalent? Why?

3.6 The metals and carbon

The metals

In a metal, the atoms are packed tightly together in a regular pattern. The tight packing causes outer electrons to get separated from their atoms, resulting in a lattice of ions in a sea of electrons. This arrangement gives metals like copper a crystal structure:

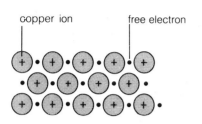

copper ion   free electron

The copper ions are held together by their attraction to the electrons between them. The strong forces of attraction are called **metallic bonds**.

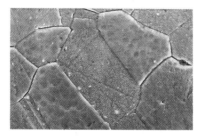

The regular arrangement of ions results in **crystals** of copper. This shows the crystals in a piece of copper magnified 1000 times.

The copper crystals are called **grains**. A lump of copper consists of millions of grains joined together. You need a microscope to see them.

Their properties   All metals are different. But they share these properties to some extent:

1 They are usually hard. (But think about sodium and mercury.)
2 They are **tough**. That is the opposite of **brittle**. When you hammer them, they will not shatter.
3 They can withstand being squashed. We say they are **strong** under compression or have high compression strength
4 They can also withstand being stretched. They are **strong under tension** or have **high tensile strength**.
5 With enough force, most metals can be bent or pressed into shape (**malleable**) or drawn into wires (**ductile**). The layers of ions can slide over each other without the metallic bonds breaking.
6 Because the atoms are closely packed, metals are usually 'heavy' or **dense**. Lead is very dense. But aluminium is light!

Lead atoms have almost eight times the mass of aluminium atoms. They are packed like this, and so lead is dense.

Although much lighter than lead atoms, aluminium atoms are not that much smaller. So the metal is much less dense.

7 They are good conductors of electricity, because the free electrons can move through the lattice carrying charge.
8 Good conductors of heat. The free electrons take in heat energy.
9 They usually have high melting points because it takes a lot of heat energy to break the strong metallic bonds. Copper melts at 1083 °C. But Group 1 metals have quite low melting points (98 °C for sodium) and mercury is a liquid at room temperature.

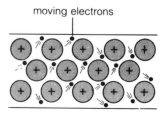

Lead is so dense it can be used to stop the passage of radiation.

moving electrons

56

Carbon

Carbon occurs naturally in two forms, diamond and graphite. Although both contain only carbon atoms they have very different properties, because of how the atoms are bonded and arranged.

Diamond Diamond is made of a lattice of carbon atoms:

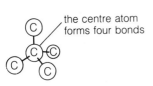
the centre atom forms four bonds

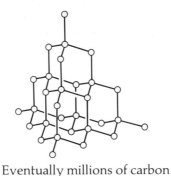

A carbon atom forms covalent bonds to *four* others, as shown above. Each outer atom then bonds to three more, and so on.

Eventually millions of carbon atoms are bonded together, in a giant covalent structure. This is part of it.

The result is a single crystal of diamond like this one. It has been cut, shaped and polished to make it sparkle.

Diamond has these properties:
1 It is very hard – the hardest substance known. This is because each atom is held in place by four strong bonds. For the same reason, it has a very high melting point (3550 °C).
2 It cannot conduct electricity, because there are no ions or free electrons in it to carry charge.

Graphite Graphite is made of flat sheets of carbon atoms:

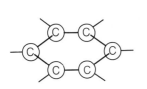

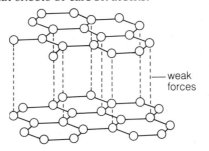
weak forces

Each carbon atom forms covalent bonds to *three* others. This gives rings of six atoms.

These join to make flat sheets that lie on top of each other, held together by weak forces.

Even at low magnification the layered structure of graphite shows up clearly.

Graphite has these properties:
1 It is soft and slippery. This is because the sheets of atoms can slide over each other easily.
2 It is a good conductor of electricity. This is because each atom has four outer electrons, but forms only three bonds. The fourth electron is free to move through the graphite, carrying charge.

Questions

1 Describe the structure of a metal.
2 Explain why metals can be drawn out into wires.
3 Why can metals conduct electricity? Would you expect a molten metal to conduct? Why?
4 Draw a diagram to show how the carbon atoms bond in: **a** diamond **b** graphite
5 Why are diamonds so hard?
6 Why is graphite soft and slippery?

Questions on Section 3

1 The table below shows the structure of several particles:

Particle	Electrons	Protons	Neutrons
A	12	12	12
B	12	12	14
C	10	12	12
D	10	8	8
E	9	9	10

a Which three particles are neutral atoms?
b Which particle is a negative ion? What is the charge on this ion?
c Which particle is a positive ion? What is the charge on this ion?
d Which two particles are isotopes?
e Use the table on page 30 to identify the particles A to E.

2 This question is about the ionic bond formed between the metal lithium (atomic number 3) and the nonmetal fluorine (atomic number 9).
a How many electrons are there in a lithium atom? Draw a diagram to show its electron structure. (You can show the nucleus as a dark circle at the centre.)
b How does a metal atom obtain a full outer shell of electrons?
c Draw the structure of a lithium ion, and write a symbol for the ion.
d How many electrons are there in a fluorine atom? Draw a diagram to show its electron structure.
e How does a nonmetal atom become a negative ion?
f Draw the structure of a fluoride ion, and write a symbol for the ion.
g Draw a diagram to show what happens when a lithium atom reacts with a fluorine atom.
h Draw the arrangement of ions in the compound that forms when lithium and fluorine react together.
i Write a name and a formula for the compound in part h.

3 Na^+ O_2 Al CH_4 N I^-
a From the list above, select:
   i two atoms
   ii two molecules
   iii two ions
b What do the following symbols represent?
   i Na^+
   ii I^-
c Name the compound made up from Na^+ and I^- ions, and write a formula for it.

4 a The electronic configuration of a neon atom is (2,8). What is special about the outer shell of a neon atom?
b The electronic configuration of a calcium atom is (2,8,8,2). What must happen to a calcium atom for it to achieve a noble gas structure?
c Draw a diagram of an oxygen atom, showing its eight protons (p), eight neutrons (n), and eight electrons (e).
d What happens to the outer-shell electrons of a calcium atom, when it reacts with an oxygen atom?
e Name the compound that is formed when calcium and oxygen react together. What type of bonding does it contain?
f Write a formula for the compound in e.

5 a Write down a formula for each of the following:
   i a nitrate ion
   ii a sulphate ion
   iii a carbonate ion
   iv a hydroxide ion
b The metal strontium forms ions with the symbol Sr^{2+}. Write down the formula for each of the following:
   i strontium oxide
   ii strontium chloride
   iii strontium nitrate
   iv strontium sulphate

6 A molecule of a certain gas can be represented by the diagram on the right.
a What is the gas? What is its formula?
b What type of bonding holds the atoms together?
c Name another compound with this type of bonding.
d What do the symbols ● and × represent?

7 Draw diagrams to show how the electrons are shared in the following molecules:
a fluorine, F_2
b water, H_2O
c methane, CH_4
d trichloromethane, $CHCl_3$
e oxygen, O_2
f hydrogen sulphide, H_2S
Draw the shapes of molecules a, b and e.

8 a An oxygen molecule is represented as $O = O$. What does the double line mean? How many electrons from each atom take part in bonding?
b A molecule of carbon dioxide (CO_2) can be drawn as $O = C = O$. Draw a diagram to show how the electrons are shared in the molecule.

9 Nitrogen is in Group 5 of the periodic table, and its atomic number is 7. It exists as molecules, each containing two atoms.

a Write the formula for nitrogen.

b What type of bonding would you expect between the two atoms?

c How many electrons does a nitrogen atom have in its outer shell?

d How many electrons must each atom obtain a share of, in order to gain a full shell of 8 electrons?

e Draw a diagram to show the bonding in a nitrogen molecule. You need show only the outer-shell electrons. (It may help you to look back at the bonding in an oxygen molecule, on page 52.)

f The bond in a nitrogen molecule is called a **triple bond**. Can you explain why?

g Nitrogen (N_2), oxygen (O_2), chlorine (Cl_2) and hydrogen (H_2) all exist as diatomic molecules. What does *diatomic* mean?

10 Arrange the following substances in three groups, according to their structure:

Group A – giant ionic structure.
Group B – giant covalent structure.
Group C – molecular structure.

methane	sulphur	sodium chloride
silica	oxygen	naphthalene
iodine	graphite	water
diamond	ethanol	copper(II) sulphate
ice	ammonia	potassium hydroxide
nitrogen	bromine	hydrogen chloride
phosphorus	steam	sulphur dioxide

11 This table gives information about properties of certain substances A to G.

Substance	Melting point (°C)	Electrical conductivity solid	Electrical conductivity liquid	Solubility in water
A	−112	poor	poor	insoluble
B	680	poor	good	soluble
C	−70	poor	poor	insoluble
D	1495	good	good	insoluble
E	610	poor	good	soluble
F	1610	poor	poor	insoluble
G	660	good	good	insoluble

a Which of the substances are metals? Give reasons for your choice.

b Which of the substances are ionic compounds? Give reasons for your choice.

c Two of the substances have very low melting points, compared with the rest. Explain why these could *not* be ionic compounds.

d Two of the substances are molecular. Which ones are they?

e Which substance is a giant covalent structure?

f Which one would you expect to be very hard?

g Which substances would you expect to be soluble in tetrachloromethane?

12 Silicon lies directly below carbon in Group 4 of the periodic table. The table below lists the melting and boiling points for silicon, carbon (diamond), and their oxides.

Substance	Symbol or formula	Melting point (°C)	Boiling point (°C)
Carbon	C	3730	4530
Silicon	Si	1410	2400
Carbon dioxide	CO_2	sublimes at −78 °C	
Silicon dioxide	SiO_2	1610	2230

a In which state are the two *elements* at room temperature (20 °C)?

b Is the structure of carbon (diamond) giant covalent or molecular?

c What type of structure would you expect silicon to have? Give reasons.

d In which state are the two oxides at room temperature?

e What type of structure does carbon dioxide have?

f Does silicon dioxide have the same structure as carbon dioxide? What is your evidence?

13 Hydrogen bromide is a compound of the two elements hydrogen and bromine. It has a melting point of −87 °C and a boiling point of −67 °C. Bromine is one of the halogens (Group 7 of the periodic table).

a Is hydrogen bromide a solid, a liquid or a gas at room temperature (20 °C)?

b Is it made of molecules, or does it have a giant structure? How can you tell?

c What type of bond is formed between the hydrogen and bromine atoms in hydrogen bromide? Show this on a diagram.

d Write a formula for hydrogen bromide.

e Name two other compounds that would have bonding similar to hydrogen bromide.

f Write formulae for these two compounds.

14 a Use the structures of diamond and graphite to explain why:

i graphite is used for the 'lead' in pencils;

ii diamonds are used in cutting tools.

b Give reasons why:

i copper is used in electrical wiring;

ii steel is used for domestic radiators.

c Ethanol is used as the solvent in perfume and aftershave, because it evaporates easily. What does that tell you about the bonding in it?

4.1 Where solids get their properties

Where does a solid get its properties?

Some solids are hard and strong. Others are soft and stretchy. Some conduct electricity. Others don't. What makes them all so different?

The properties of a solid depend on three things:

1 What particles it is made of.
2 How the particles are bonded together.
3 How they are arranged.

For example:

Lead is heavy because its atoms are heavy and tightly packed. It conducts electricity because of its free electrons. It can be rolled into sheets because the layers of ions slide over each other.

Diamond is light because carbon atoms are light. It is hard and strong because each atom forms a strong covalent bond to four others, giving a rigid lattice.

Rubber is made of long molecules, mostly carbon atoms. It is flexible because there is no rigid lattice. The molecules are coiled but they stretch when you pull.

In general, if a solid:

● conducts electricity it must be a metal (unless it is graphite!).
● is hard, rigid and strong and doesn't conduct electricity even when melted, it must have a criss-cross lattice of covalent bonds.
● is flexible and doesn't conduct electricity, it must consist of separate molecules held together by forces of attraction between them.
● is elastic, its molecules must be able to coil and uncoil.

On the following pages, you can look at some everyday solids and see if they fit these rules!

Different properties for different purposes.

What properties should these materials have?

Designing materials to meet a need

These days scientists design materials to give the properties they want! **Alloys** are a good example. They are usually a mixture of metals:

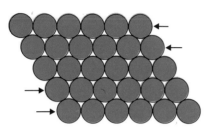 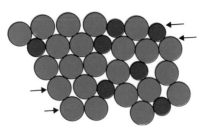 

In aluminium the atoms are packed like this. The atoms are light so the metal is light. The bonds are strong but the layers can slide over each other easily.

Copper atoms are smaller. When copper is mixed with aluminium, the layers cannot slide over each other so easily. So the mixture is stronger.

The alloy obtained by mixing 4% copper with 96% aluminium is called duralium. It is so light and strong that it is used for aircraft parts.

Testing materials

Scientists must test a new material rigorously to make sure it behaves as they want. For example, a new alloy for aeroplane bodies must be light. But it must also be strong under tension so that it can withstand an explosion, and strong under compression in case there's a crash. These are the kinds of test that are carried out:

  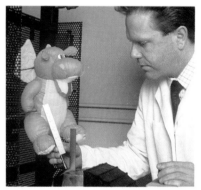

Compression testing.

Tension testing (at high temperatures).

Preparing a child's toy for a flammability test.

Questions

1 Write down the three things that affect the properties of a solid.
2 Look at the rules on the opposite page. Now say what you expect the bonding to be like in:
   a a kitchen tile
   b chewing gum
   c a plastic ruler
3 Write a list of all the solids in the photo at the bottom left of the opposite page. Beside each, write down what kind of bonding you think it has.
4 What is an alloy? Why are alloys made?
5 Write down a list of tests you could do to find out as much as possible about the properties of the polythene in a plastic bag.

4.2 Glasses and ceramics

Glasses

There are several kinds of glass, but the main ingredient of all of them is **sand**. The most common glass around the house is **soda-lime** glass. It is made by heating sand, limestone (calcium carbonate), sodium carbonate and recycled glass in a furnace to around 1300 °C.

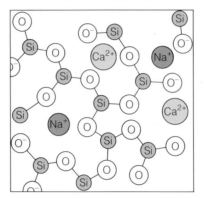

In the furnace the carbonates break down to give a molten mixture of oxides. On cooling, this forms a giant irregular structure made of silicon and oxygen atoms . . .

. . . with sodium and calcium ions trapped inside. This shows a cross-section through it. Each silicon atom is covalently bonded to *four* oxygen atoms, but one is hidden from view.

While still soft, the glass can be moulded and blown into shape to make items such as bottles, jars and drinking glasses. Or it can be cooled as flat sheets, for making windows.

Ceramics

Bricks, tiles, mugs and clay flowerpots are all **ceramics**. Ceramics are made from clay, which is a mixture of silicates and other minerals dug from the ground. For example, it usually contains kaolinate, $Al_2Si_2O_5(OH)_4$, in the form of tiny flat crystals.

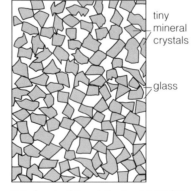

tiny mineral crystals

glass

Wet clay is easy to shape because the tiny crystals in it slide over each other. When it dries out, it keeps its new shape because the crystals stick together.

It is heated in a kiln to 1000 °C or more. This causes a series of reactions that produce other minerals and glass. The result is lots of tiny mineral crystals bonded together by glass.

Now the object is usually covered in glaze and fired again. Glaze is a suspension of minerals in water. Reactions in the glaze make the surface waterproof.

Scrap glass which will be added to the furnace along with the raw materials.

These dishes were fired in a furnace lined with fireclay, a ceramic with a very high melting point.

The properties of glasses and ceramics

Since ceramics are tiny mineral crystals bonded by glass, you might expect glasses and ceramics to share some properties – and they do:

1 Both are hard and will not bend, because of the strong covalent bonds in the giant structure.
2 Because the bonding is covalent, both are electrical insulators. (But note that *molten* glass can conduct electricity because the ions in it are free to move.)
3 They conduct heat, though not as well as metals do.
4 Both can withstand being squashed without crumbling: they are strong under compression. Think of bricks!
5 But they cannot withstand being stretched. They are weak under tension and will crack.
6 They will shatter when dropped: they are brittle. Their structures are irregular (unlike diamond) and weaker at some points than others. Energy from an impact spreads through the structure causing it to break at the weak points.
7 For the same reason, a sudden temperature change may also shatter them.
8 They are unreactive and won't corrode, since they already consist mainly of unreactive oxides.
9 They have very high melting points, and will melt over a range of temperatures since they are not pure substances.

Ceramics are used as electrical insulators.

But there are two important differences between them:

● Glasses are usually transparent; ceramics are not.
● Glasses have a lower melting point than ceramics.

Questions

1 Suggest reasons why:
  a glass is used for fish tanks;
  b glass shatters when you drop it;
  c soda-lime glass is not very expensive;
  d recycled glass is used in making new glass.

2 Suggest reasons why:
  a glasses and ceramics are alike in many ways;
  b crowns for teeth are often made from ceramics;
  c spaceships are covered with thin ceramic tiles;
  d ceramics are used to support cables on pylons.

4.3 Plastics

You probably own quite a lot of things made of plastic: pen, toothbrush and comb included! All plastics are carbon compounds. Most of them are made from chemicals obtained from crude oil.

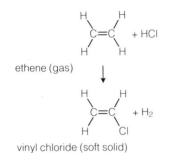

ethene (gas)

vinyl chloride (soft solid)

Ethene is made from compounds in crude oil. It reacts with hydrochloric acid to give chloroethene or **vinyl chloride**. When this is mixed with warm water under pressure . . .

. . . the double bonds break and the molecules join to form chains thousands of atoms long. The result is a rigid solid called poly(chloroethene) or **polyvinyl chloride** or **PVC**.

It is mixed with pigments to give it colour and with plasticizers to make it flexible. It is used for window frames, wellies, credit cards, curtain rails, raincoats and so on . . .

There are many other types of plastic. For example:

Name	Used for
Polythene	plastic bags and bottles, dustbins, 'cling film'
Polystyrene	plastic cups, fast-food cartons, packaging foam
Nylon	rope, bristle for brushes, tights, carpet
Melamine	'unbreakable' dishes and mugs, ashtrays
Urea-formaldehyde	electric plugs and switches
Phenol-formaldehyde	door handles, saucepan handles

Plastics have these properties:

1. They are all carbon compounds.
2. Their molecules are long chains of atoms, made by joining lots of small molecules together. So they are called **polymers**. *Poly* means *many*. They are produced by a reaction called **polymerization**.
3. Because they are molecular compounds, they do not conduct electricity. They are electrical and thermal insulators.
4. They are unreactive. Most are not affected by water, or oxygen, or other chemicals, which makes them very useful. But it also means they are difficult to dispose of. Unlike wood or paper, they do not rot away. (But **biodegradable** plastics will rot away.)
5. Some catch fire easily. When they burn, they may give out harmful gases. For example, PVC gives out choking fumes of hydrogen chloride when it burns.
6. Plastics contain mainly carbon and hydrogen atoms, which are light. So plastics are also light. But they are also strong because their molecules are so long. The longer the molecules, the larger the force of attraction between them.
   (As you will see on the next page, the molecules in some plastics are cross-linked, which makes them even stronger.)

a pair of molecules

as molecules get longer....

force of attraction between them increases

Thermoplastics and thermosets

Plastics are divided into two groups, **thermoplastics** and **thermosets**.

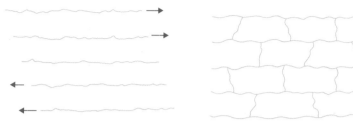

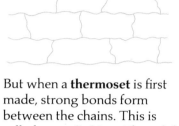

When you heat a **thermoplastic** it gets soft and runny: the heat overcomes the forces between the chains so they can slide over each other. On cooling, it hardens into its new shape.

But when a **thermoset** is first made, strong bonds form between the chains. This is called **cross-linking**. The solid sets into shape and stays hard, even when you heat it.

This shows a polythene bowl and a melamine bowl which were put in an oven for just a few seconds. Which is a thermoplastic and which is a thermoset?

Thermosets   Because of the cross-linking, thermosets:

- will char or break down at high temperatures, rather than melt.
- are rigid and will break rather than bend. Dropping them may cause the giant irregular structure to break at its weak points.
- are moulded into shape *while they are being made*, because the shape can't be changed later.

Thermoplastics   Because there is no cross-linking, thermoplastics:

- are not brittle, and can be very flexible.
- stretch under tension, because the molecules slide over each other.
- melt at quite low temperatures.
- can be moulded into shape *after* they are made. So they are usually produced as powder or granules.

Moulding thermoplastics   These are some of the methods used:

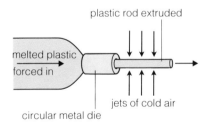

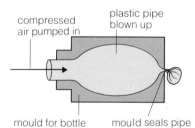

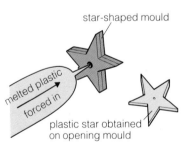

Extrusion. Melted plastic is forced through a shaped nozzle or die, then cooled to harden it. Used for things like pipes and curtain rails.

Blow moulding. Plastic pipe is clamped into a mould, then blown up using compressed air until it fits the mould. Used for plastic bottles.

Injection moulding. Melted plastic is injected into a mould at high speed, then allowed to cool. Used for things like toys, furniture, bottle caps.

Questions

1 Write down three facts which are true of all plastics.
2 Draw a diagram to show the structure of:
   a a thermoplastic   b a thermoset

3 Explain why a plastic shopping bag:
   a can take a heavy load even though it's thin;
   b may stretch out of shape;   c may cause pollution.

4.4 Synthetic fibres

A **fibre** is any substance that can be turned into a fabric by spinning and weaving or knitting it. Wool, cotton, hair, silk and the rough hairy covering on coconuts are all **natural fibres**. There are also many man-made or **synthetic** fibres, based on plastics.

These are the main groups of synthetic fibres:

polyamides such as nylon and Enkalon;
polyesters such as Terylene and Dacron;
acrylics such as Courtelle and Dralon;
polyurethanes such as Lycra and Spandex.

The non-chemical names above are the ones that manufacturers have made up. But often the chemical name is used on clothing labels. Check out the labels on your clothes!

No matter which plastic you start with, synthetic fibres are all made in much the same way:

Lycra is very stretchy!

The sticky plastic liquid is forced through tiny holes called **spinnerets** to make fine threads or **filaments**. These are dried in air. Each is made up . . .

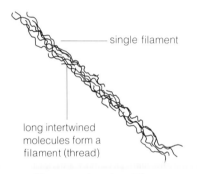

single filament

long intertwined molecules form a filament (thread)

. . . of long molecules. The filaments are stretched to line up the molecules, and wound on to rollers. Then they are twisted together to make yarn.

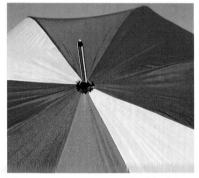

The yarn is woven or knitted to make fabric for clothing, tents, parachutes, tablecloths, umbrellas, carpets, raincoats and so on.

Synthetic fibres have these properties:

1 Like all plastics, they are carbon compounds made up of long molecules: they are polymers.
2 The longer the molecules in the filaments and the more aligned they are, the stronger the fibre will be. (This is because attractive forces between molecules increase as molecules get longer and as they get closer together.)
3 They have greater tensile strength than plastics because the molecules are more aligned.
4 But if you pull a fibre hard enough, you overcome the attractive forces between molecules. They slide apart and the fibre breaks.
5 They are smooth, so fabrics made from them wash and dry well. These are often called 'drip-dry'.
6 They usually melt at quite low temperatures: they are **thermoplastics**. You have to be careful when you iron synthetic fabrics.

The properties of a fabric depend on the fibre, and also on how it is woven or knitted. Would you expect this fabric to stretch to fit? Why?

Some unusual synthetic fibres

Viscose   Most synthetic fibres are made from chemicals obtained from oil. But viscose is made from **wood pulp**.
Wood contains cellulose, a natural polymer. The wood pulp is soaked in sodium hydroxide and other chemicals to give a sticky cellulose gum, which is then turned into fibres in the usual way. The names *viscose*, *acetate*, *tri-acetate* and *rayon* on labels all show that the fibre was made from wood pulp.

Carbon fibre   Carbon fibre is made by heating a synthetic fibre, such as viscose, to a very high temperature in the absence of air. The fibre doesn't burn. It chars away until what's left is just the carbon. As the data on the right shows, carbon fibre is very strong! It is used to make carbon-fibre-reinforced plastic for tennis rackets and golf clubs.

Glass fibre   Glass fibre is made by running molten glass into a spinner, which is a steel bowl with lots of tiny holes in it. It spins round very fast so that the molten glass flies out through the holes as thin fibres. These are sprayed with a resin to bind them together, then dried and hardened in an oven.
Glass fibre is used for insulating lofts and boilers. It can also be woven to make curtains and other furnishing fabrics.

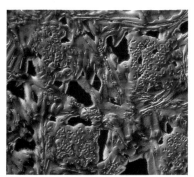

Carbon fibres are so strong they are used to make the fan blades in jet engines.

Tensile strength in N/m²

Aluminium	92
Copper	214
Steel	414
Glass fibre	1104
Carbon fibre	2760

Monofilaments

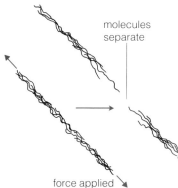

molecules separate

force applied

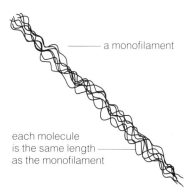

a monofilament

each molecule is the same length as the monofilament

In an ordinary synthetic fibre the bonds *within* the molecules are very strong. But if you pull hard enough, the fibre breaks because the molecules separate.

In a **monofilament**, each molecule in the fibre is a continuous chain *the same length as the fibre*. So these fibres won't break easily.

Monofilaments can be woven into fabric that is so strong it can trap speeding bullets. Good for things like bullet-proof vests!

Some plastics can be made into monofilaments. So can carbon fibre. The carbon fibre in tennis rackets and golf clubs is in the form of a monofilament embedded in plastic resin. (See page 69.)

Questions

1 Explain why a synthetic fibre can have greater tensile strength than a strip cut from a plastic bag.
2 If you pull a synthetic fibre, it may stretch a bit before it breaks. Why?
3 Viscose is quite a cheap fibre. Why?
4 How is carbon fibre made?
5 How is glass fibre made?
6 What is a *monofilament*?

4.5 Composite materials

Often one material is combined with another to give a **composite** material with better properties. For example if you melt plastic film between sheets of glass, you get **safety glass**, which is shatterproof.

Wood, bone, ivory and teeth are natural composite materials. Synthetic composite materials include concrete, reinforced concrete, glass-fibre-reinforced plastic or fibreglass, and plywood.

Bone

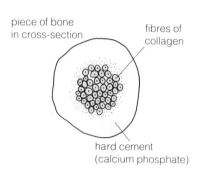

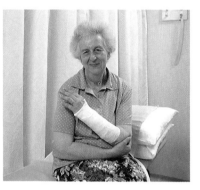

Bone is tough, flexible fibres of a protein called collagen, embedded in a hard cement of calcium phosphate. The cement contains the bone's living cells.

The result is hard, light, and slightly flexible, quite strong under tension and compression, supporting your flesh and whatever else you carry.

But calcium is continually lost from bones. If it's not absorbed again from food, the cement gets fragile and the bones break more easily.

Concrete

Concrete consists of **aggregate** (a mixture of stone chips and sand) bound together by **cement**. Cement is a mixture of calcium aluminate and calcium silicate, made by heating limestone with clay. For concrete, the cement is mixed with aggregate and water added.

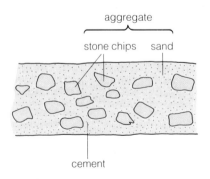

As the concrete 'sets', particles in the cement bond together forming a hard mass of crystals with water molecules trapped inside them. This is called **water of crystallization**.

The hard crystals and chips make the concrete strong under compression. But it is weak under tension – a concrete beam will crack if you place a heavy load on it.

To improve the tensile strength of concrete, steel rods or steel mesh is embedded in it. The result is called **reinforced concrete**. It is used for bridges and multi-storey buildings.

Reinforced plastic

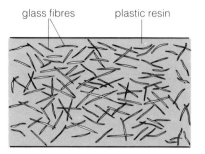

glass fibres   plastic resin

Glass-fibre-reinforced plastic is made by mixing glass fibres with a liquid plastic resin that sets hard. The result is . . .

. . . hard, strong, light, resistant to corrosion and attack by chemicals, and does not shatter easily.

Glass-fibre-reinforced plastic is used for making boats, surfboards and some car bodies.

Carbon fibres can be used instead, to give **carbon-fibre-reinforced plastic**. This is used for tennis rackets and golf clubs.

Composite fabrics

Most clothes these days are made of composite fabrics. These combine the hard-wearing and easy-care properties of synthetic fabrics with the texture and warmth of natural fabrics.

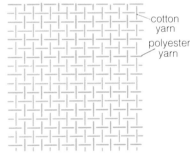

cotton yarn
polyester yarn

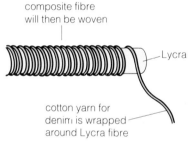

composite fibre will then be woven

Lycra

cotton yarn for denim is wrapped around Lycra fibre

Cotton fabric is usually light and comfortable, but not very hard-wearing. It wrinkles when you wash it. Ironing cotton garments can be a big chore.

Cotton can be woven with polyester to give polyester cotton. This is easier to wash and dry, wears better and needs less ironing. Good for shirts.

Denim is a tough cotton fabric. But combining the fibre with Lycra gives a tougher fabric with stretch. It keeps its shape well. Good for jeans.

Laminated materials

In some composite materials, the different materials are combined in sheets. These are called **laminates**. One example is safety glass. Another is **plywood**, where layers of different woods are glued together with the grain at right angles. In **plasterboard**, a layer of plaster is sandwiched between sheets of very strong paper. The cover of this book is a laminate of paper and a thin film of plastic.

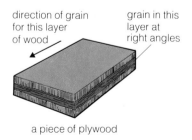

direction of grain for this layer of wood   grain in this layer at right angles

a piece of plywood

Questions

1 What is a composite material? Give two examples.
2 Explain why bone is strong under tension.
3 Would you expect plywood to bend easily? Why?
4 Name six composite materials used in your home.

Questions on Section 4

1 Many polymers can be drawn into threads which have good strength and resistance to wear and tear. List the properties you would expect from a fibre to be used for:
   a rope for mountain climbing
   b tights
   c tennis racket strings
   d underwear

2 List A below contains materials. List B contains properties.

A	B
brass	brittle and transparent
nylon	good electrical conductor
glass	brittle and opaque
ceramic	strong under compression
concrete	melts when heated, used as a fibre

   a Match each material from list A with one correct property from list B.
   b Which of the materials in list A is:
      i a composite material? ii a plastic? iii an alloy?

3 This question is about the properties of three different materials: A, B and C.

Material	Physical properties	Chemical properties

A	hard
	strong when
	   compressed
	weak when
	   stretched
	transparent
		chemically
		unreactive

B	hard strong flexible good conductor of heat and electricity	corrodes in air reacts with acids

C	flexible
	poor conductor of
	   heat and electricity
	easily melted and
	   moulded
		burns on heating in
		   air, with poisonous
		   fumes

   a Which material could you use for:
      i a bottle for storing sulphuric acid?
      ii a washing-up bowl?
      iii reinforcing concrete?
   b Which material could be:
      i glass? ii plastic? iii metal?
   c Draw up a similar list of properties for:
      i porcelain ii concrete

4 a What are the raw materials for soda-lime glass?
   b Draw a diagram showing how the atoms in soda-lime glass are arranged.
   c Make a list of the properties of glass by choosing from the words below.
   hard flexible electrical conductor
   transparent unreactive weak under tension
   regular structure high compressive strength
   d Which of the properties you have chosen for glass are also shared by ceramics? Tick them on your list.

5 Copy the following sentences about plastics, and complete them by inserting the correct word from this list:
   thermosets thermoplastics monomers
   polymerization polymers chains
   Plastics are made from molecules containing long .... of atoms called .... They are made by a process called .... in which .... join together. Plastics like nylon melt on heating and are called .... while resins char when heated and are examples of ....

6 You are a designer of equipment for outdoor expeditions. Part of your job is to choose the right materials. What properties would you require for:
   a the frame of a rucksack?
   b the shoulder straps of a rucksack?
   c the padding in an anorak?
   d the soles of walking boots?
   e the cover for a map?
   f a water bottle?

7 The following words describe the properties of materials:
   hard insulator flexible corrosive elastic
   transparent biodegradable light strong
   a Write these properties out as a list, and beside each write the opposite property.
   b Which are the important properties that make plastics a good replacement for wood in the home?
   c Which properties in the list above do *not* apply to ceramics?
   d i How does an elastic material behave?
      ii What type of structure would you expect an elastic material to have?

8 Explain why each of the listed plastics is used for the job described.
   a Perspex is used for car windscreens.
   b Polypropylene is used for buckets and bowls.
   c PVC is used for 'plastic' macs.
   d Polytetrafluoroethylene (PTFE) is used for coating frying pans.
   e Expanded polystyrene is used for ceiling tiles.
   f Polyester is used for making photographic film.

9 Soft drinks can be sold in glass or plastic bottles.

glass   plastic

a What properties do both materials share, which make them suitable for fizzy drinks?
b Suggest three advantages of making the bottles from plastic.
c From the point of raw materials, why might it be better to use glass?
d Why are there more bottle banks for recycling glass than there are for recycling plastics?
e Glass has a much higher melting point than a thermoplastic. Explain why, using a diagram of their structures to help you.

10 Mortar is a composite material used to join bricks together to make walls. It is made by adding water to a mixture of sand and cement. The strength of the mortar depends on the proportion of sand to cement used in the mixture (for example 1:1, 2:1, 3:1 and so on).
You are to describe an investigation that could be carried out to find the effect of composition on the strength of the mortar. Below is the equipment and material you will use. The mortar will be placed in ice trays and allowed to set into cubes for testing.

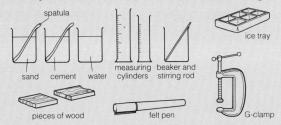

spatula

ice tray

sand   cement   water   measuring   beaker and
                            cylinders   stirring rod

pieces of wood   felt pen   G-clamp

a Why is mortar described as a *composite material*?
b Why is thorough mixing of the sand and cement important?
c What else is added to make the mortar?
d You have a G-clamp and pieces of wood. Describe how you could compare the strengths under compression for the samples of mortar. What could you count?
e Which variable will you change in the experiment?
f What must you keep the same, to ensure the test is fair? List as many things as you can.
g Why is it important to test the mortar for strength under compression, but not important to test it for strength under tension?
h Name another composite material where it would be essential to test strength under tension.
i Apart from strength, what other properties must mortar have to make it useful as a building material?

11 Glass-fibre-reinforced plastic (GRP) is widely used for making surfboards, golf clubs and boats.
a Why is plastic on its own not used for making surfboards?
b Explain why adding glass fibre produces a material very suitable for the objects above.

12 Carbon fibre in the form of a monofilament is embedded in a polymer resin to make frames for tennis rackets. Compared with wood or aluminium, the carbon fibre reduces the amount of energy transferred from the collision with the ball to the player's elbow.
a Wood and aluminium have both been used for tennis rackets. Make two lists showing the properties that make them suitable.
b The weight-to-strength ratio for carbon-fibre rackets is much higher than for those made from wood or aluminium. Explain why this is.
c Tennis players are now less likely to suffer from tennis elbow. Why do you think this is?

13 Laminated windscreens are made by bonding a sheet of a clear thermoplastic between two sheets of glass.
a Explain the term *laminated*.
b Draw a diagram to show a cross-section of a laminated windscreen.
c Suggest a simple method for bonding a sheet of a thermoplastic between two sheets of glass.
d Why does glass crack when struck by an object?
e What happens to the energy of impact?
f Explain how the addition of the plastic sheet can prevent a crack spreading across a windscreen, when the windscreen is struck.
g Explain why a laminated windscreen breaks into tiny pieces which cling together.
h Why is a laminated windscreen to be preferred to one made from perspex?

14 When a ceramic is made, the atoms on the surface react together to form a giant lattice where all the bonds are strong. Below this hard surface layer the ceramic contains tiny holes or pores.
a Why are ceramics able to withstand temperatures of over 1500 °C without melting?
b Name one use of ceramics where this property is exploited.
c What is the main disadvantage of ceramic materials?
d Explain how their structure gives rise to this property.
e How might the properties of a ceramic change, if it was combined with a fibre to make a composite material?
f Invent a material, combining a ceramic and a fibre. Suggest how it could be made. What could it be used for?

5.1 The masses of atoms

Relative atomic mass

A single atom weighs hardly anything. For example, the mass of a single hydrogen atom is only about 0.000 000 000 000 000 000 000 002 grams. Very small numbers like that are awkward to use, so scientists had to find a simpler way to express the mass of an atom. Here is what they did.

First, they chose a carbon atom to be the standard atom.

 — an atom of $^{12}_{6}C$

The mass spectrometer was invented by a British scientist called Aston, in 1919.

Next, they fixed its mass as exactly 12 units. (It has 6 protons and 6 neutrons. They ignored the electrons.)
Then they compared all the other atoms with this standard atom, using a machine called a mass spectrometer, and found values for their masses, like this:

This is the standard atom, $^{12}_{6}C$. Its mass is exactly 12.	This magnesium atom is twice as heavy as the standard atom, so its mass must be 24.	This hydrogen atom is $\frac{1}{12}$ as heavy as the standard atom, so its mass must be 1.

The mass of an atom found by comparing it with the $^{12}_{6}C$ atom is called its **relative atomic mass**, or **RAM** for short.
So the RAM of hydrogen is 1 and the RAM of magnesium is 24.

RAMs and isotopes   Not all atoms of an element are exactly the same. For example, when scientists examined chlorine in the mass spectrometer, they found there were two types of chlorine atom:

 — this one has a mass of 35

 — this one has a mass of 37

These atoms are the **isotopes** of chlorine. (They have different masses because one has two neutrons more than the other.)
It was found that out of every four chlorine atoms, three have a mass of 35 and one has a mass of 37. Using this information, the *average* mass of a chlorine atom was calculated to be 35.5.
Most elements have more than one isotope, and these have to be taken into account when finding RAMs:
The RAM of an element is the average mass of its isotopes compared with an atom of $^{12}_{6}C$.
For most elements, the RAMs work out very close to whole numbers. They are usually rounded off to whole numbers, to make calculations easier.

To calculate the average mass of a chlorine atom, first find the total mass of four atoms:

$$3 \times 35 = 105$$
$$1 \times 37 = \underline{37}$$
$$142$$

The average mass $= \dfrac{142}{4}$

$$= 35.5$$

The RAMs of some common elements Here is a list of them:

Element	Symbol	RAM	Element	Symbol	RAM
Hydrogen	H	1	Chlorine	Cl	35.5
Carbon	C	12	Potassium	K	39
Nitrogen	N	14	Calcium	Ca	40
Oxygen	O	16	Iron	Fe	56
Sodium	Na	23	Copper	Cu	64
Magnesium	Mg	24	Zinc	Zn	65
Sulphur	S	32	Iodine	I	127

Finding the mass of an ion:

Mass of sodium atom = 23, so mass of sodium ion = 23, since a sodium ion is just a sodium atom minus an electron, and an electron has hardly any mass. **An ion has the same mass as the atom from which it is made.**

Formula mass

Using a list of RAMs, it is easy to work out the mass of any molecule or group of ions. Check these examples using the information above:

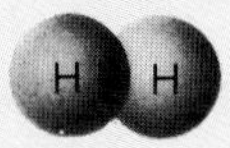

Hydrogen gas is made of molecules. Each molecule contains 2 hydrogen atoms, so its mass is 2. ($2 \times 1 = 2$)	The formula of water is H_2O. A water molecule contains 2 hydrogen atoms and 1 oxygen atom, so its mass is 18. ($2 \times 1 + 16 = 18$)	Sodium chloride (NaCl) contains a sodium ion for every chloride ion. The mass of a 'unit' of sodium chloride is 58.5. ($23 + 35.5 = 58.5$)

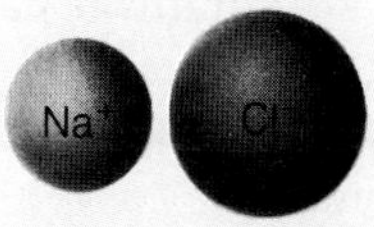

The mass of a substance found in this way is called its **formula mass**, because it can be obtained by adding up the masses of the atoms in the formula. If the substance is made of molecules, its mass can also be called the **relative molecular mass**, or **RMM**.
So the RMM of hydrogen is 2 and the RMM of water is 18.
Here are some more examples:

Substance	Formula	Atoms in formula	RAM of atoms	Formula mass
Nitrogen	N_2	2N	N = 14	$2 \times 14 = \mathbf{28}$
Ammonia	NH_3	1N	N = 14	$1 \times 14 = 14$
		3H	H = 1	$3 \times 1\ = \underline{\ 3}$
				Total $= \underline{\mathbf{17}}$
Magnesium nitrate	$Mg(NO_3)_2$	1Mg	Mg = 24	$1 \times 24 = \ \ 24$
		2N	N = 14	$2 \times 14 = \ \ 28$
		6O	O = 16	$6 \times 16 = \underline{\ \ 96}$
				Total $= \underline{\mathbf{148}}$

Questions

1 What is the relative atomic mass of an element?
2 What is the RAM of the iodide ion, I^-?
3 Show that the formula mass for chlorine (Cl_2) is 71.
4 What is the RMM of butane, C_4H_{10}?

5 Work out the formula mass of:
  a oxygen, O_2 b iodine, I_2
  c methane, CH_4 d ethanol, C_2H_5OH
  e ammonium sulphate $(NH_4)_2SO_4$

5.2 The mole

What is a mole?

On the last two pages you read about relative atomic masses and formula masses. These are not just boring numbers – they are very important for a chemist to know.

If you work out the RAM or formula mass of a substance, and then weigh out that number of grams of the substance, you can say how many atoms or molecules it contains.

This is very useful, since single atoms and molecules are far too small to be seen or counted.

For example, the RAM of carbon is 12. The photograph on the right shows 12 grams of carbon. The heap contains 602 000 000 000 000 000 000 000 carbon atoms.

This is called a **mole** of atoms.
The number is called **Avogadro's number**, after Avogadro, an Italian scientist who lived over a hundred years ago.
It is usually written in a shorter way as 6.02×10^{23}.
(The 10^{23} shows that you must move the decimal point 23 places to the right to get the full number.)
Now look at these:

Sodium is made up	Iodine is made up	Water is made up
of single sodium atoms. Its RAM is **23**.	of iodine molecules. Its formula is I_2. Its formula mass is **254**.	of water molecules. Its formula is H_2O. Its formula mass is **18**.
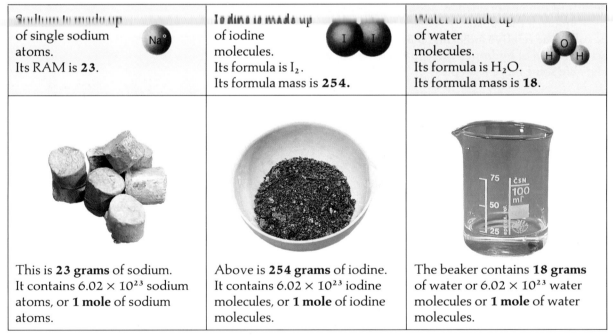		
This is **23 grams** of sodium. It contains 6.02×10^{23} sodium atoms, or **1 mole** of sodium atoms.	Above is **254 grams** of iodine. It contains 6.02×10^{23} iodine molecules, or **1 mole** of iodine molecules.	The beaker contains **18 grams** of water or 6.02×10^{23} water molecules or **1 mole** of water molecules.

From these examples you should see that:
One mole of a substance is 6.02×10^{23} particles of the substance. It is obtained by weighing out the RAM or formula mass, in grams.

Finding the mass of a mole

You can find the mass of 1 mole of any substance, by these steps:

1 Write down the symbol or formula of the substance.
2 Find out its RAM or formula mass.
3 Express that mass in grams.

This table shows more examples:

Substance	Symbol or formula	RAMs	Formula mass	Mass of 1 mole
Helium	He	He = 4	4	**4 grams**
Oxygen	O_2	O = 16	$2 \times 16 = 32$	**32 grams**
Ethanol	C_2H_5OH	C = 12 H = 1 O = 16	$2 \times 12 = 24$ $6 \times 1 = 6$ $1 \times 16 = \underline{16}$ $\underline{46}$	**46 grams**

Some calculations on the mole

Example 1 Calculate the mass of:
a 0.5 moles of bromine atoms.
b 0.5 moles of bromine molecules.

a The RAM of bromine is 80, so 1 mole of bromine *atoms* has a mass of 80 grams. Therefore 0.5 moles of bromine *atoms* has a mass of 0.5×80 grams, or 40 grams.

b A bromine *molecule* contains 2 atoms, so its formula mass is 160. Therefore 0.5 moles of bromine *molecules* has a mass of 0.5×160 grams, or 80 grams.

Formula mass or RMM = 160

So, to find the mass of a given number of moles:
mass = mass of 1 mole × number of moles

Example 2 How many moles of oxygen molecules are there in 64 grams of oxygen, O_2?

The formula mass of oxygen gas is 32, so 32 grams of it is 1 mole.

Therefore 64 grams is $\frac{64}{32}$ moles, or 2 moles.

Formula mass or RMM = 32

So, to find the number of moles in a given mass:

$$\text{number of moles} = \frac{\text{mass}}{\text{mass of 1 mole}}$$

Questions

1 How many atoms are in 1 mole of atoms?
2 How many molecules are in 1 mole of molecules?
3 What name is given to the number 6.02×10^{23}?
4 Find the mass of 1 mole of:
  a hydrogen atoms   **b** iodine atoms
  c chlorine atoms   **d** chlorine molecules

5 Find the mass of 2 moles of:
  a oxygen atoms   **b** oxygen molecules
6 Find the mass of 3 moles of ethanol, C_2H_5OH.
7 How many moles of molecules are there in:
  a 18 grams of hydrogen, H_2?
  b 54 grams of water?

5.3 Compounds and solutions

The percentage of an element in a compound

Methane is a compound of carbon and hydrogen, with the formula CH_4. The mass of a carbon atom is 12, and the mass of each hydrogen atom is 1, so the formula mass of methane is 16.

RAMs:
$C = 12, H = 1$
Formula mass $= 16$

You can find what fraction of the total mass is carbon, and what fraction is hydrogen, like this:

Mass of carbon as fraction of total mass $= \dfrac{\text{mass of carbon}}{\text{total mass}} = \dfrac{12}{16}$ or $\dfrac{3}{4}$

Mass of hydrogen as fraction of total mass $= \dfrac{\text{mass of hydrogen}}{\text{total mass}}$

$$= \dfrac{4}{16} \text{ or } \dfrac{1}{4}$$

These fractions are usually written as percentages. To change a fraction to a percentage, you just multiply it by 100:

$\dfrac{3}{4} \times 100 = \dfrac{300}{4} = 75$ per cent or 75% $\dfrac{1}{4} \times 100 = \dfrac{100}{4} = 25\%$

So 75% of the mass of methane is carbon, and 25% is hydrogen. We say that the **percentage composition** of methane is 75% carbon, 25% hydrogen.

The famous scientist, John Dalton and a pupil, collecting methane from a pond. Methane also occurs in oil wells and natural gas wells. No matter where it comes from, its composition is always the same.

Calculations on composition

Here is how to calculate the percentage of an element in a compound:
1 Write down the formula of the compound.
2 Using a list of RAMs, work out its formula mass.
3 Write the mass of the element you want, as a fraction of the total.
4 Multiply the fraction by 100, to give a percentage.

Example   Fertilizers contain nitrogen, which plants need to help them grow. One important fertilizer is ammonium nitrate, NH_4NO_3. What is the percentage of nitrogen in it?

The formula mass of the compound is 80, as shown on the right. The element we are interested in is nitrogen.

Mass of nitrogen in the formula $= 28$

Mass of nitrogen as fraction of total $= \dfrac{28}{80}$

Mass of nitrogen as percentage of total $= \dfrac{28}{80} \times 100 = \mathbf{35\%}$

Finding the formula mass for ammonium nitrate:

The RAMs are: $N = 14$, $H = 1$, $O = 16$.
The formula contains 2 N, 4 H and 3 O, so the formula mass is:

$2\,N = 2 \times 14 = 28$
$4\,H = 4 \times 1 = 4$
$3\,O = 3 \times 16 = \underline{48}$
 Total $= \underline{80}$

The concentration of a solution

Below are three solutions of copper(II) sulphate in water.
Each has the same volume, 1 dm³. (**1 dm³** is **1 litre**, or **1000 cm³**.)

A

B

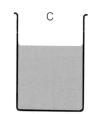
C

Only 2.5 grams of copper(II) sulphate was dissolved to make this solution. It is **dilute**.

This one contains 25 grams of the compound. It is more **concentrated** than A.

This one contains 250 grams of the compound, so it is more concentrated than either A or B.

Each solution above has a different **concentration**.
The concentration of a solution is the amount of solute, in grams or moles, that is dissolved in 1 dm³ of solution.

The concentrations of the copper(II) sulphate solutions in grams are:
    A – 2.5 grams per dm³
    B – 25 grams per dm³
    C – 250 grams per dm³

Finding the concentration in moles

The formula mass of copper(II) sulphate must first be worked out. The answer is 250, as shown on the right. Therefore 1 mole of the compound has a mass of 250 grams.

The formula of copper(II) sulphate is $CuSO_4.5H_2O$. The formula contains 1 Cu, 1 S, 9 O, and 10 H, so the formula mass is:

$$\begin{aligned} 1\,Cu &= 1 \times 64 = 64 \\ 1\,S &= 1 \times 32 = 32 \\ 9\,O &= 9 \times 16 = 144 \\ 10\,H &= 10 \times 1 = \underline{10} \\ Total &= \underline{250} \end{aligned}$$

Solution A   It has 2.5 grams of the compound in 1 dm³ of solution.

$$2.5 \text{ grams} = \frac{2.5}{250} \text{ moles} = 0.01 \text{ moles}$$

so its concentration is **0.01 moles per dm³**.
Mole per dm³ is often shortened just to M, so the concentration of solution A can be written as **0.01 M**.

Solution C   It has 250 grams of the compound in 1 dm³ of solution.
    250 grams = 1 mole
    so its concentration is **1 mole per dm³**, or **1 M** for short.
A solution that contains 1 mole of solute per dm³ is often called a **molar solution**, so C is a molar solution.

Questions

1 In sulphur dioxide, SO_2, what is the percentage of:
   a sulphur? **b** oxygen?
   The RAMs are: S = 32, O = 16.
2 Find the percentage of hydrogen and oxygen in ammonium nitrate, NH_4NO_3.
3 Find the percentage of hydrogen in water, H_2O.

4 Work out the concentration of solution B above in moles per dm³.
5 The formula mass of sodium hydroxide is 40. How many grams of sodium hydroxide are there:
   a in 1 litre of a molar solution?
   b in 500 cm³ of a 1 M solution?

5.4 Formulae of compounds

What a formula tells you

The formula of carbon dioxide is **CO₂**. Some molecules of it are shown on the right. You can see that:

| 1 carbon atom | combines with | 2 oxygen atoms |

. It follows that

| 1 mole of carbon atoms | combines with | 2 moles of oxygen atoms |

.

RAMs: C = 12, O = 16

Moles can be changed to grams, using RAMs. So we can write:

| 12 g of carbon | combines with | 32 g of oxygen |

.

This means that 6 g of carbon combines with 16 g of oxygen, and so on. Therefore, from the formula of a compound, you can tell:
- **how many moles of the different atoms combine.**
- **how many grams of the different elements combine.**

Now look at the formula of ammonia, **NH₃**. It shows that:

| 1 mole of nitrogen atoms (14 grams) | combines with | 3 moles of hydrogen atoms (3 grams) |

Do you agree? Check the masses using the RAMs on the right.

RAMs: H = 1, N = 14

Finding formulae from masses

From the formula of a compound, you can tell what masses of the elements will combine. But you can also do things the other way round. Starting with the masses that combine, you can work out the formula of the compound. These are the steps:

| Start with the number of **grams** that combine. | → | Change the grams to **moles of atoms**. | → | This gives you the **ratio** in which atoms **combine**. | → | So you can write down a **formula**. |

A formula found by this method is called an **empirical formula**.

Example 1   32 grams of sulphur combines with 32 grams of oxygen, to form the compound sulphur dioxide. What is its formula?

First, change the masses to moles of atoms.
The RAM of sulphur is 32, and the RAM of oxygen is 16, so:

$\frac{32}{32}$ moles of sulphur atoms combines with $\frac{32}{16}$ moles of oxygen atoms, or

1 mole of sulphur atoms combines with 2 moles of oxygen atoms, so
1 sulphur atom combines with 2 oxygen atoms.
The formula of sulphur dioxide is therefore **SO₂**.

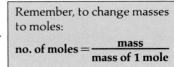

> Remember, to change masses to moles:
>
> **no. of moles** $= \dfrac{\textbf{mass}}{\textbf{mass of 1 mole}}$

78

Example 2   20 grams of calcium reacts completely with 19 grams of fluorine, to form calcium fluoride. What is its formula?

The RAMs are: Ca = 40, F = 19. Changing masses to moles:

$\frac{20}{40}$ moles of calcium atoms reacts with $\frac{19}{19}$ moles of fluorine atoms, or

0.5 moles of calcium atoms reacts with 1 mole of fluorine atoms, so
1 mole of calcium atoms reacts with 2 moles of fluorine atoms.
The formula of calcium fluoride is therefore **CaF$_2$**.

Finding masses by experiment

To work out a formula, you first need to know what masses of the elements combine. This must be found out by experiment.
For example, magnesium combines with oxygen like this:

magnesium + oxygen $\longrightarrow$ magnesium oxide

and the masses that combine can be found like this:

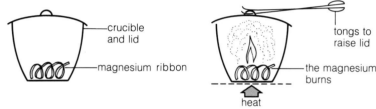

First, a crucible and lid is weighed. Next a coil of magnesium ribbon is added, and the new mass found.

Then the crucible is heated, and the lid carefully raised a little, at intervals, to let in oxygen. The magnesium burns brightly.

When burning is complete, the crucible is allowed to cool, with its lid still on.
Then it is weighed again.

The results   Here are some sample results, and the calculation:
Mass of crucible + lid   = 25.2 g
Mass of crucible + lid + magnesium   = 27.6 g
Mass of crucible + lid + magnesium oxide = 29.2 g

Mass of magnesium   = 27.6 g − 25.2 g = 2.4 g
Mass of magnesium oxide = 29.2 g − 25.2 g = 4.0 g
Mass of oxygen therefore =   4.0 g −   2.4 g = 1.6 g

So 2.4 g of magnesium combines with 1.6 g of oxygen. The formula of magnesium oxide can now be found.
The RAMs are: Mg = 24, O = 16. Changing the masses to moles:

$\frac{2.4}{24}$ moles of magnesium atoms combines with $\frac{1.6}{16}$ moles of oxygen atoms.

0.1 moles of magnesium atoms combines with 0.1 moles of oxygen atoms.
1 mole of magnesium atoms combines with 1 mole of oxygen atoms.
The formula of magnesium oxide is therefore **MgO**.

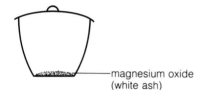

Questions

1 **a** How many moles of carbon atoms combine with 4 moles of hydrogen atoms, to form methane, CH$_4$?
b How many grams of hydrogen combine with 12 grams of carbon, to form methane?

2 How would you change *grams* to *moles of atoms*?

3 To form iron(II) sulphide, 56 g of iron combines with 32 g of sulphur. Find its formula.
(The RAMs are: Fe = 56, S = 32.)

4 In hydrogen sulphide, there is 1 g of hydrogen for every 16 g of sulphur. Find its formula.

Questions on Section 5

Relative atomic masses are given on page 250. Use the approximate values given in the table.

1 How many grams are there in:
  a 1 mole of copper atoms?
  b 1.5 moles of sulphur atoms?
  c 2 moles of magnesium atoms?
  d 5 moles of carbon atoms?
  e 10 moles of chlorine atoms?
  f 0.1 moles of nitrogen atoms?
  g 0.2 moles of neon atoms?
  h 0.6 moles of hydrogen atoms?
  i 1.5 moles of oxygen atoms?

2 How many grams are there in:
  a 1 mole of hydrogen molecules, H_2?
  b 2 moles of hydrogen molecules, H_2?
  c 1 mole of oxygen molecules, O_2?
  d 0.5 moles of chlorine molecules, Cl_2?
  e 2 moles of phosphorus molecules, P_4?
  f 4 moles of sulphur molecules, S_8?

3 Find how many moles of atoms there are, in:
  a 32 g of sulphur
  b 48 g of magnesium
  c 23 g of sodium
  d 14 g of lithium
  e 1.4 g of lithium
  f 3.1 g of phosphorus
  g 6.4 g of oxygen
  h 5.4 g of aluminium
  i 2 g of hydrogen
  j 0.6 g of carbon

4 For each pair, decide which of the two substances contains the greater number of atoms.
  a 80 g of sulphur, 80 g of calcium
  b 80 g of sulphur, 80 g of oxygen
  c 1 mole of sulphur atoms, 8 moles of chlorine atoms
  d 1 mole of sulphur atoms, 1 mole of oxygen molecules
  e 4 moles of sulphur atoms, $\frac{1}{8}$ mole of sulphur molecules (S_8)

5 For each pair, decide which of the two solutions contains the greater number of moles of solute.
  a 1 litre of 1 M sodium chloride (NaCl), 1 litre of 2 M sodium chloride
  b 500 cm³ of 1 M sodium chloride, 1 litre of 1 M sodium chloride
  c 1 litre of 0.1 M sodium chloride, 100 cm³ of 2 M sodium chloride
  d 250 cm³ of 2 M sodium chloride, 1 litre of 1 M sodium hydroxide (NaOH)
  e 20 cm³ of 0.5 M sodium chloride, 40 cm³ of 1 M sodium chloride

6 How many grams are there in:
  a 1 mole of water, H_2O?
  b 5 moles of water?
  c 1 mole of anhydrous copper(II) sulphate, $CuSO_4$?
  d 1 mole of hydrated copper(II) sulphate, $CuSO_4 . 5H_2O$?
  e 2 moles of ammonia, NH_3?
  f 0.5 moles of ammonium carbonate, $(NH_4)_2CO_3$?
  g 0.3 moles of calcium carbonate, $CaCO_3$?
  h $\frac{1}{5}$ mole of magnesium oxide, MgO?
  i 0.1 moles of sodium thiosulphate, $Na_2S_2O_3$
  j 2 moles of iron(III) chloride, $FeCl_3$

7 1 mole of sodium carbonate (Na_2CO_3) contains 2 moles of sodium atoms, 1 mole of carbon atoms and 3 moles of oxygen atoms.
  In the same way, write down the number of moles of each atom present in 1 mole of:
  a lead oxide, Pb_3O_4
  b ammonium nitrate, NH_4NO_3
  c calcium hydroxide, $Ca(OH)_2$
  d dinitrogen tetroxide, N_2O_4
  e ethanol, C_2H_5OH
  f ethanoic acid, CH_3COOH
  g hydrated iron(II) sulphate, $FeSO_4 . 7H_2O$
  h iron(III) ammonium sulphate, $NH_4Fe(SO_4)_2$

8 The formula of calcium oxide is CaO. The RAMs are: Ca = 40, O = 16.
  Complete the following statements:
  a 1 mole of Ca (...... g) and 1 mole of O (...... g) combine to form ...... mole of CaO (...... g).
  b 4.0 g of calcium and ...... g of oxygen combine to form ...... g of calcium oxide.
  c When 0.4 g of calcium reacts with oxygen, the increase in mass is ...... g.
  d If 6 moles of CaO were decomposed to calcium and oxygen, ... moles of Ca and ...... moles of O_2 would be obtained.
  e The percentage by mass of calcium in calcium oxide is ......%.

9 Two samples of copper oxide were made by different methods. The oxides were then converted to copper. The following results were obtained:

	Sample 1	Sample 2
Mass of copper oxide	16.0 g	32.0 g
Mass of copper obtained	12.8 g	25.6 g

  a Calculate the percentage composition of copper in each sample.
  b Were the two oxides of the same or different composition?

10 In a reaction to make manganese from manganese oxide, the following results were obtained: 174 g of manganese oxide produced 110 g of manganese. (RAMs: Mn = 55, O = 16)

a What mass of oxygen is there in 174 g of manganese oxide?

b How many moles of oxygen atoms is this?

c How many moles of manganese atoms are there in 110 g of manganese?

d What is the simplest formula of manganese oxide?

11 The following results were obtained in an experiment to find the formula of magnesium oxide:

Mass of crucible = 12.5 g
Mass of crucible + magnesium = 14.9 g
Mass of crucible + magnesium oxide = 16.5 g
(The RAMs are: Mg = 24, O = 16)

a What mass of magnesium was used in the experiment?

b How many moles of magnesium atoms is this?

c What mass of oxygen combined with the magnesium?

d How many moles of oxygen atoms is this?

e Use your answers to parts **b** and **d** to find the formula of magnesium oxide.

12 27 g of aluminium burns in a stream of chlorine to form 133.5 g of aluminium chloride.
(RAMs: Al = 27, Cl = 35.5)

a What mass of chlorine is present in 133.5 g of aluminium chloride?

b How many moles of chlorine atoms is this?

c How many moles of aluminium atoms are present in 27 g of aluminium?

d Use your answers for parts **b** and **c** to find the simplest formula of aluminium chloride.

e 133.5 g of aluminium chloride is dissolved to form 1 dm³ of an aqueous solution. What is the concentration in moles per dm³ of this aluminium chloride solution? (1 dm³ is the same as 1 litre.)

f 1 dm³ of an aqueous solution is made using 13.35 g of aluminium chloride. What is its concentration in moles per dm³?

13 Ammonia (NH_3), carbon dioxide (CO_2), hydrogen (H_2), and oxygen (O_2) are all gases at room temperature.

a Which of the gases are compounds?

b Which are diatomic?

c For each gas, calculate the mass of one mole of molecules.

d How many moles of ammonia molecules are there in 34 g of ammonia?

e How many moles of hydrogen atoms are there in 34 g of ammonia?

f What mass of carbon dioxide contains the same number of oxygen atoms as 16 g of oxygen?

14 An oxide of copper can be converted to copper by heating it in a stream of hydrogen in the apparatus shown below:

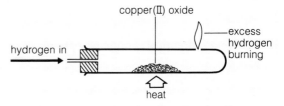

The hydrogen supply was turned on, and hydrogen was allowed to pass through the test-tube for some time before the excess gas was lit. The test-tube was heated until all the copper oxide was converted to copper. The apparatus was allowed to cool, with hydrogen still passing through, before it was dismantled.

a Copy and complete the word equation for the reaction:

copper oxide + hydrogen → copper + ...

b Why was the hydrogen allowed to pass through the apparatus for some time, before the excess was lit? Why was the excess hydrogen burned?

The experiment was repeated several times, by different groups in the class. Each group used a different mass of oxide. The results are shown below.

Group	Mass of copper oxide/g	Mass of copper produced/g	Mass of oxygen lost/g
1	0.62	0.55	0.07
2	0.90	0.80	0.10
3	1.12	1.00	0.12
4	1.69	1.50	
5	1.80	1.60	

c Work out the missing figures for the table.

d On graph paper, plot the mass of copper against the mass of oxygen. (Show the mass of copper along the x-axis and the mass of oxygen along the y-axis.) Then draw the best straight line through the origin and the set of points.

e From the graph, find the mass of oxygen which would combine with 1.28 g of copper.

f Calculate the mass of oxygen which would combine with 128 g of copper.
(Remember, $1.28 \times 100 = 128$)

g How many moles of copper atoms are there in 128 g of copper? (Cu = 64)

h How many moles of oxygen atoms combine with 128 g of copper? (O = 16)

i What is the simplest formula of this oxide of copper?

j What is another name for the *simplest formula* found in this way?

6.1 Physical and chemical change

A substance can be changed by heating it, adding water to it, mixing another substance with it, and so on. The change that takes place will be either a **chemical** change or a **physical** one.

Chemical change

mixture of iron filings and sulphur

iron filings cling to the magnet
magnet

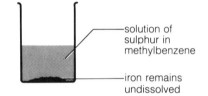
solution of sulphur in methylbenzene

iron remains undissolved

1 Some yellow sulphur and black iron filings are mixed together.

2 The mixture is easily separated again, by using a magnet to attract the iron . . .

3 . . . or by dissolving the sulphur in methylbenzene (a solvent).

the contents glow even after the Bunsen is removed

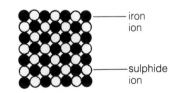
black solid
magnet

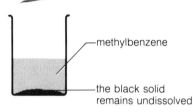
methylbenzene

the black solid remains undissolved

4 But when the mixture is *heated*, it glows brightly. The yellow specks of sulphur disappear. A black solid forms.

5 This black solid is not at all like the mixture. It is not affected by a magnet . . .

6 . . . and none of it dissolves in methylbenzene.

The black solid is obviously a new chemical substance.
When it produces a new chemical substance, a change is called a chemical change.
So in step 4, a chemical change has taken place. The iron and sulphur have **reacted** together, to form the compound iron sulphide.

The difference between the mixture and the compound In the mixture above, iron and sulphur particles are mixed closely together, but they are not bonded to each other – the iron particles still behave like iron, and the sulphur particles like sulphur. During the reaction, however, iron and sulphur atoms form ions which bond to each other. The magnet and solvent now have no effect.

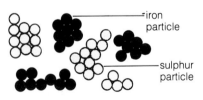
iron particle

sulphur particle

iron ion

sulphide ion

Mixture Iron and sulphur particles mixed together. Each particle contains many atoms.

Compound Iron ions and sulphide ions bonded together to form iron sulphide.

The compound iron sulphide occurs in the earth as **iron pyrites**.

The signs of a chemical change

A chemical change is usually called a **chemical reaction**. You can tell when a chemical reaction has taken place, by these signs:

1 One or more new chemical substances are formed.
The new substances usually look quite different from the starting substances. For example:

iron   +   sulphur   ⟶   iron sulphide
(black filings)   (yellow powder)   (black solid)

2 Energy is taken in or given out, during the reaction.
In step 4, a little energy in the form of heat from a Bunsen is needed to start off the reaction between iron and sulphur. But the reaction gives out heat once it begins.
A reaction that gives out heat energy is **exothermic**.
A reaction that takes in heat energy is **endothermic**.
So the reaction between iron and sulphur is exothermic. The reactions that take place when you fry an egg are endothermic. Energy can be given out in the form of light or sound, as well as heat. For example magnesium burns in air with a bright white light and a hiss.

3 The change is usually difficult to reverse.
You would need to carry out several other reactions, to get back the iron and sulphur from iron sulphide.

Fireworks contain magnesium and other substances. When they burn, the reactions give out energy in the form of heat, light, and sound.

Physical change

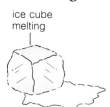

ice cube melting

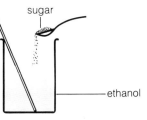

sugar

ethanol

Ice turns to water at 0 °C. It is easy to change the water back to ice again, by cooling it.

Sugar dissolves in ethanol. You can separate the two again by distilling the solution.

The reactions that take place when you fry an egg are endothermic.

No new chemical substances are formed in these changes.
For example, although ice and water *look* different, they are both made of water molecules, and have the formula H_2O.
When no new chemical substance is formed, a change is called a physical change.
So the changes above are both physical changes. Physical changes are usually easy to reverse.

Questions

1 Explain the difference between a *mixture* of iron and sulphur and the *compound* iron sulphide.
2 What are the signs of a chemical change?
3 What is: **a** an exothermic reaction?
   b an endothermic reaction?

4 Is the change chemical or physical? Give reasons.
   a Glass bottle breaking.
   b Butter and sugar being made into toffee.
   c Wool being knitted into a sweater.
   d Coal burning in air.

6.2 Equations for chemical reactions

The reaction between carbon and oxygen When carbon is heated in oxygen, they react together and carbon dioxide is formed. The carbon and oxygen are called **reactants**, because they react together. Carbon dioxide is the **product** of the reaction. You could show the reaction by a diagram, like this:

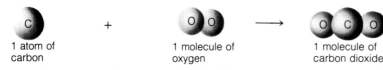

or in a shorter way, by using symbols, like this:

$$C \quad + \quad O_2 \quad \longrightarrow \quad CO_2$$

This short way to describe the reaction is called a **chemical equation**.

The reaction between hydrogen and oxygen When hydrogen and oxygen react together, the product is water. The diagram is:

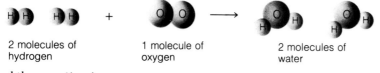

and the equation is:

$$2H_2 \quad + \quad O_2 \quad \longrightarrow \quad 2H_2O$$

Can you see why there is a 2 in front of H_2 and H_2O, in the equation? Now look at the number of atoms on each side of the equation.

On the left:	On the right:
4 hydrogen atoms	4 hydrogen atoms
2 oxygen atoms	2 oxygen atoms

The numbers of hydrogen and oxygen atoms are the same on both sides of the equation. This is because atoms do not *disappear* during a reaction – they are just *rearranged*, as shown in the diagram.
When the numbers of different atoms are the same on both sides, an equation is said to be **balanced**. An equation which is not balanced is not correct. Check the equation for the reaction between carbon and oxygen above. Is it balanced?

Adding more information to equations Reactants and products may be solids, liquids, gases or solutions. You can show their states by adding **state symbols** to the equations. The state symbols are:

   (s) for solid   (l)   for liquid
   (g) for gas   (aq) for aqueous solution (solution in water)

For the two reactions above, the equations with state symbols are:

$$C\,(s) \ + \ O_2\,(g) \quad \longrightarrow \quad CO_2\,(g)$$
$$2H_2\,(g) \ + \ O_2\,(g) \quad \longrightarrow \quad 2H_2O\,(l)$$

In a coal fire, the main reaction is $C + O_2 \rightarrow CO_2$

The reaction between hydrogen and oxygen gives out so much energy that it is used to power rockets. The hydrogen and oxygen are carried as liquids in the fuel tanks.

How to write the equation for a reaction

These are the steps to follow, when writing an equation:

1 Write the equation in words.
2 Now write the equation using symbols. Make sure all the formulae are correct.
3 Check that the equation is balanced, for each type of atom in turn. *Make sure you do not change any formulae.*
4 Add the state symbols.

Example 1   Calcium burns in chlorine to form calcium chloride, a solid. Write an equation for the reaction, using the steps above.

1 Calcium + chlorine $\longrightarrow$ calcium chloride
2   Ca   + Cl_2 $\longrightarrow$ $CaCl_2$
3 Ca: 1 atom on the left and 1 atom on the right.
   Cl: 2 atoms on the left and 2 atoms on the right.
   The equation is balanced.
4 $Ca\,(s)$ + $Cl_2\,(g)$ $\longrightarrow$ $CaCl_2\,(s)$

Example 2   In industry, hydrogen chloride is formed by burning hydrogen in chlorine. Write an equation for the reaction.

1 Hydrogen + chlorine $\longrightarrow$ hydrogen chloride
2   H_2 + Cl_2 $\longrightarrow$ HCl
3 H: 2 atoms on the left and 1 atom on the right.
   Cl: 2 atoms on the left and 1 atom on the right.
   The equation is *not* balanced. It needs another molecule of hydrogen chloride on the right. So a 2 is put *in front of* the HCl.
   H_2 + Cl_2 $\longrightarrow$ 2HCl
   The equation is now balanced. Do you agree?
4 $H_2\,(g)$ + $Cl_2\,(g)$ $\longrightarrow$ $2HCl\,(g)$

Example 3   Magnesium burns in oxygen to form magnesium oxide, a white solid. Write an equation for the reaction.

1 Magnesium + oxygen $\longrightarrow$ magnesium oxide
2   Mg   + O_2 $\longrightarrow$ MgO
3 Mg: 1 atom on the left and 1 atom on the right.
   O: 2 atoms on the left and 1 atom on the right.
   The equation is *not* balanced. Try this:
   Mg + O_2 $\longrightarrow$ 2MgO (Note, the 2 goes *in front of* the MgO.)
   Another magnesium atom is now needed on the left:
   2Mg + O_2 $\longrightarrow$ 2MgO
   The equation is balanced.
4 $2Mg\,(s)$ + $O_2\,(g)$ $\longrightarrow$ $2MgO\,(s)$

Magnesium burning in oxygen.

Questions

1 What do + and $\longrightarrow$ mean, in an equation?
2 Balance the following equations:
   a $Na\,(s)$ + $Cl_2\,(g)$ $\longrightarrow$ $NaCl\,(s)$
   b $H_2\,(g)$ + $I_2\,(g)$ $\longrightarrow$ $HI\,(g)$
   c $Na\,(s)$ + $H_2O\,(l)$ $\longrightarrow$ $NaOH\,(aq)$ + $H_2\,(g)$
   d $NH_3\,(g)$ $\longrightarrow$ $N_2\,(g)$ + $H_2\,(g)$
   e $C\,(s)$ + $CO_2\,(g)$ $\longrightarrow$ $CO\,(g)$
   f $Al\,(s)$ + $O_2\,(g)$ $\longrightarrow$ $Al_2O_3\,(s)$
3 Aluminium burns in chlorine to form aluminium chloride, $AlCl_3\,(s)$. Write an equation for the reaction.

6.3 Calculations from equations

What an equation tells you

When carbon burns in oxygen, the equation for the reaction is:

$$C\,(s)\ +\ O_2\,(g)\ \longrightarrow\ CO_2\,(g)$$

This equation tells you that:

| 1 carbon atom | reacts with | 1 molecule of oxygen | to give | 1 molecule of carbon dioxide |

Now suppose there was 1 *mole* of carbon atoms. These would react with 1 *mole* of oxygen molecules:

| 1 mole of carbon atoms | reacts with | 1 mole of oxygen molecules | to give | 1 mole of carbon dioxide molecules |

Moles can be changed to grams, using RAMs and formula masses. The RAMs are: C = 12, O = 16. So the formula mass of CO_2 is: (12 + 16 + 16) = 44, and we can write:

| 12 g of carbon | reacts with | 32 g of oxygen | to give | 44 g of carbon dioxide |

You can find out the same kind of information from any equation:
The equation for a reaction tells you how many moles and how many grams of each substance take part in the reaction.

But that's not all . . .

The reaction above involves gases. Thanks to Avogadro and other scientists, we know this fact about gases:
At room temperature and pressure, one mole of any gas has a volume of 24 dm³ (24 litres or 24 000 cm³).
24 dm³ is the **molar gas volume** at room temperature and pressure **(rtp)**. So from the equation above you can tell that:

| 12 g of carbon | reacts with | 24 dm³ of oxygen | to give | 24 dm³ of carbon dioxide |

(The gas volumes are measured at rtp.)

Does the mass change, during a reaction?

Now look what happens to the total mass, during the above reaction:

Mass of carbon and oxygen at the start: 12 g + 32 g = **44 g**
Mass of carbon dioxide at the end: **44 g**

The total mass has not changed, during the reaction. This is because

the atoms taking part have not changed either.
During a chemical reaction atoms do not disappear and new atoms do not form. The atoms just get rearranged.

Calculations from equations

Example 1 Hydrogen burns in oxygen to form water. The equation for the reaction is: $2H_2(g) + O_2(g) \longrightarrow 2H_2O(l)$ How much oxygen is needed to burn 1 gram of hydrogen?

1 The RAMs are: $H = 1$, $O = 16$. So $H_2 = 2$ and $O_2 = 32$.
2 $2Hg(g) + O_2(g) \longrightarrow 2H_2O(l)$
   2 moles of hydrogen molecules needs 1 mole of oxygen molecules
   4 g of hydrogen needs 32 g of oxygen (moles changed to grams)
   1 g of hydrogen needs 8 g of oxygen
3 The reaction needs **8 g** of oxygen.

These models show how the atoms are rearranged during the reaction between hydrogen and water. The equation is $2H_2 + O_2 \rightarrow 2H_2O$.

Example 2 The equation for the reaction between iron and sulphur is: $Fe(s) + S(s) \longrightarrow FeS(s)$. When 7 g of iron is heated with excess sulphur, how much iron(II) sulphide is formed? (*Excess* sulphur means *more than enough* sulphur for the reaction.)

1 The RAMs are: $Fe = 56$, $S = 32$. So $FeS = 56 + 32 = 88$.
2 $Fe(s) + S(s) \longrightarrow FeS(s)$
   1 mole of iron atoms gives 1 mole of iron sulphide units so
   56 g of iron gives 88 g of iron sulphide

   1 g of iron gives $\frac{88}{56}$ g of iron sulphide

   7 g of iron gives $7 \times \frac{88}{56}$ g of iron sulphide

3 $7 \times \frac{88}{56} = 11$ so **11 g** of iron(II) sulphide is produced.

Example 3 What volume of hydrogen will react with 24 dm³ of oxygen to form water? (Gas volumes measured at rtp.)

1 $2H_2(g) + O_2(g) \longrightarrow 2H_2O(l)$
2 2 moles of hydrogen gas reacts with 1 mole of oxygen gas.
   1 mole of any gas has a volume of 24 dm³ at rtp, so
   2×24 dm³ of hydrogen will react with 24 dm³ of oxygen.
3 **48 dm³** of hydrogen is needed.

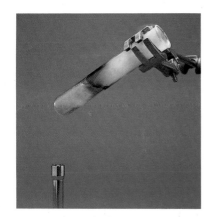

The reaction between iron and sulphur. It is an exothermic reaction but heat is needed to get it started.

Questions

1 Explain why the total mass does not change during a reaction.

2 The reaction between magnesium and oxygen is:
   $2Mg(s) + O_2(g) \longrightarrow 2MgO(s)$
   a Write a word equation for the reaction.
   b How many moles of magnesium atoms react with 1 mole of oxygen molecules?
   c The RAMs are: $Mg = 24$, $O = 16$.
   How many grams of oxygen react with:
      i 48 g of magnesium?   **ii** 12 g of magnesium?

3 Copper(II) carbonate breaks down on heating:
   $CuCO_3(s) \xrightarrow{heat} CuO(s) + CO_2(g)$
   a Write a word equation for the reaction.
   b Find the mass of 1 mole of each substance in the reaction. ($Cu = 64$, $C = 12$, $O = 16$)
   c i If 31 g of copper(II) carbonate is used, how many grams of carbon dioxide will form?
      ii How many moles of carbon dioxide is this?
      iii What volume will it occupy at rtp?

6.4 Different types of chemical reaction (I)

Thousands of reactions go on every day in laboratories, factories, traffic jams, central heating systems, the school kitchen, the atmosphere, the earth we walk on, and within our own bodies. But it is not quite as confusing as you might think, because these reactions can be divided into quite a small number of different types.

Combination or synthesis

Often two or more substances react together to form *just one product*. This type of reaction is called a **combination** or **synthesis**. The reaction between iron and sulphur (page 83) is an example:

iron + sulphur $\longrightarrow$ iron sulphide
$Fe(s) + S(s) \longrightarrow FeS(s)$

Decomposition

Some reactions have *just one reactant*, which breaks down to give two or more simpler products. This is called **decomposition**. An example is the decomposition of calcium carbonate or **limestone**:

calcium carbonate $\xrightarrow{\text{heat}}$ calcium oxide + carbon dioxide
(limestone)   (quicklime)

$$CaCO_3(s) \xrightarrow{\text{heat}} CaO(s) + CO_2(g)$$

The label on the arrow shows that heat is needed for this reaction. Decomposition caused by heat is called **thermal decomposition**.

Decomposition can also be caused by **light**. For example silver chloride, a white solid, breaks down in light to give tiny black crystals of silver.

In cement works, limestone is decomposed to quicklime and used to make cement. The decomposition is carried out in rotating kilns.

silver chloride $\xrightarrow{\text{light}}$ silver + chlorine

$$2AgCl(s) \xrightarrow{\text{light}} 2Ag(s) + Cl_2(g)$$

Silver bromide and silver iodide decompose in the same way. These reactions are important in photography. Photographic film and paper have a coating of silver chloride, bromide, or iodide in gelatine. Where light strikes it, the silver compound decomposes, giving a dark image. The rest is washed away during processing.

Electrolysis

Decomposition can also be brought about by electricity. This process is called **electrolysis**. For example if you connect two graphite rods to a battery and stand them in molten lithium chloride, the lithium chloride decomposes. Beads of silvery lithium metal form at one rod and bubbles of green chlorine gas at the other:

lithium chloride $\xrightarrow{\text{electricity}}$ lithium + chlorine

$$2LiCl(l) \xrightarrow{\text{electricity}} 2Li(l) + Cl_2(g)$$

You can find out more about electrolysis in Section 7 (page 100).

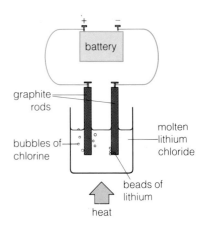

Fermentation

Fermentation is a special kind of decomposition brought about by natural organisms such as yeast. Yeast breaks down glucose, a sugar found in fruit, vegetables and cereals such as barley. The reaction produces **alcohol** and carbon dioxide:

glucose $\xrightarrow{\text{yeast}}$ alcohol (ethanol) | carbon dioxide

$C_6H_{12}O_6\,(aq) \xrightarrow{\text{yeast}} 2C_2H_5OH\,(aq) + 2CO_2\,(g)$

It also produces energy, which the yeast cells use to multiply. This reaction is the basis for making wine and brewing beer.

Precipitation

Sometimes when two solutions are mixed together, they react to give an insoluble product. The product appears as a suspension or **precipitate**, and the reaction is called a **precipitation**.

For example when aqueous solutions of sodium chloride and silver nitrate are mixed, a white precipitate of silver chloride forms:

$AgNO_3\,(aq) + NaCl\,(aq) \longrightarrow AgCl\,(s) + NaNO_3\,(aq)$

Sodium nitrate is soluble in water, so it remains in solution. The silver chloride precipitates because it is insoluble. Soon the precipitate turns black. Can you explain why?

Precipitation is useful in many ways. It allows you to:

- produce new materials. It is widely used in the paint industry for making **pigments** which are insoluble in water.
- remove unwanted ions from solution.
- identify ions in a solution. For example if you think a solution contains lead ions, you could add a few drops of sodium chloride solution. If lead ions are present, lead chloride will precipitate. Unlike silver chloride, it does not go black on standing in light.

silver nitrate solution

white precipitate

sodium chloride solution

The pigments in these paints were made by precipitation. But suppose pigments were soluble. What problems might that cause us?

Neutralization

When acids react with substances called **bases**, their acidity is destroyed. They are **neutralized**, and the reaction is called a **neutralization**. It always produces a salt and water. For example:

copper(II) oxide + sulphuric acid $\longrightarrow$ copper(II) sulphate + water

$CuO\,(s) + H_2SO_4\,(aq) \longrightarrow CuSO_4\,(aq) + H_2O\,(l)$

You can find out more about neutralization of bases on page 133.

Questions

1 Give one example of: **a** a combination reaction **b** thermal decomposition
2 When you take a photo you make use of a decomposition reaction. Explain how.
3 What is electrolysis? Give an example.
4 What is fermentation? Write an equation for it.
5 Explain why precipitation can be used to get rid of unwanted ions.

6.5 Different types of chemical reaction (II)

Oxidation and reduction

In many reactions oxygen is added to or *removed* from a substance. For example when hydrogen is passed over black copper(II) oxide in the apparatus below, the black powder turns pink:

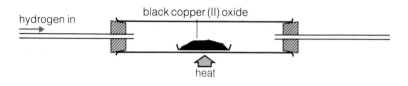

This reaction is taking place:

copper(II) oxide + hydrogen $\longrightarrow$ copper + water
$$CuO\,(s) + H_2\,(g) \longrightarrow Cu\,(s) + H_2O\,(g)$$

The copper(II) oxide is losing oxygen. It is being **reduced**.
The hydrogen is gaining oxygen. It is being **oxidized**.
If a substance loses oxygen during a reaction, it is reduced.
If a substance gains oxygen during a reaction, it is oxidized.
Reduction and oxidation always take place together in a reaction.

In the reaction above, copper(II) oxide gets reduced because hydrogen takes its oxygen away. So hydrogen is the **reducing agent**. Or you could say that hydrogen gets oxidized because copper(II) oxide gives it oxygen. So copper(II) oxide is the **oxidizing agent**.

Reduction is important in industry. For example many metals occur in the earth as oxides, or as other compounds that can easily be changed into oxides. The oxides are reduced to get the pure metal. You can find out more about this on page 152.

Rusting Rusting is the name we give to the oxidation of iron or steel in damp air. It is also called **corrosion**:

iron + oxygen + water vapour $\longrightarrow$ iron(III) hydroxide (rust)
$$4Fe\,(s) + 3O_2\,(g) + 6H_2O\,(g) \longrightarrow 4Fe(OH)_3\,(s)$$

This is one example of an unwelcome oxidation. Rust weakens structures such as car bodies, iron railings, lampposts and ships' hulls, and shortens their useful life. Preventing it can cost a lot of money. You can find out more about it on page 160.

Rancidity Oxidation can also cause things to 'go off' in the kitchen. For example, butter, margarine and cooking oils all contain **unsaturated fats**, which are fats containing carbon atoms with double bonds between them. The double bonds are unstable so oxygen reacts with them, giving substances with unpleasant flavours and smells. The oil, or butter, or margarine, turns **rancid**.

Manufacturers sometimes add **anti-oxidants** to fats, oils, and fatty foods to prevent oxidation taking place. You can also slow down oxidation by storing the foods in a cool dark place.

Making iron from its ore. Iron ore is mainly iron(III) oxide. It is reduced to iron in the blast furnace. Molten iron runs out from the bottom of the furnace.

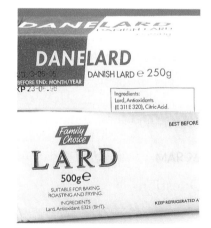

Why is it a good idea to keep lard in the fridge?

Combustion

Combustion is often just called **burning**. It usually means the reaction of a substance with oxygen from the air. During it:

a The substance that burns is oxidized to produce oxides.

b Energy is given out in the form of heat, light or sound.

Here are some examples of combustion.

1 The combustion of magnesium in the lab:

$$\text{magnesium} + \text{oxygen} \longrightarrow \text{magnesium oxide}$$
$$2Mg\,(s) + O_2\,(g) \longrightarrow 2MgO\,(s)$$

The reaction gives out heat, brilliant white light and a fizzy noise.

2 The combustion of North Sea gas in a central heating system. North Sea gas is mainly **methane**. It burns like this:

$$\text{methane} + \text{oxygen} \longrightarrow \text{carbon dioxide} + \text{water vapour}$$
$$CH_4\,(g) + 2O_2\,(g) \longrightarrow CO_2\,(g) + 2H_2O\,(g)$$

The reaction gives out a little light and sound in the boiler, and a great deal of heat which is carried around the house. Note that this time *two* different oxides are produced! They are both waste gases and are carried out into the air through the boiler flue.

3 The combustion of glucose in your body. During digestion food is broken down into simpler substances. For example, the carbohydrates in rice, potatoes and bread are broken down to form **glucose**. This reacts with oxygen in your body cells:

$$\text{glucose} + \text{oxygen} \longrightarrow \text{carbon dioxide} + \text{water vapour}$$
$$C_6H_{12}O_6\,(aq) + 6O_2\,(g) \longrightarrow 6CO_2\,(g) + 6H_2O\,(g)$$

This combustion reaction has a special name: **respiration**. It gives out the energy you need for warmth, breathing, moving and everything else you do. Compare it with the combustion of North Sea gas above. What do you notice? How are the waste gases removed this time?

Help, fire! Combustion is not always so welcome. You might not be too happy if your house burned down! A fire needs three things to keep going: fuel, oxygen (air), and heat. So, to stop a fire:

1 Cut off the fuel. Turn off the gas or electricity. Cover puddles of oil or petrol with sand or soil.

2 Get rid of the heat. Cool things down with water. But don't use water on burning petrol or oil because they will just float and spread further. And don't use it on burning electrical appliances because it can conduct electricity and give you a shock.

3 Cut off the air supply. Cover burning things with foam, carbon dioxide or a fire blanket. But don't use foam on electrical appliances because it too can conduct and give shocks.

Combustion of petrol is what makes cars move. A mixture of petrol and air is burned in the engine.

Controlled combustion in safe conditions.

With the air supply cut off, the fat fire will go out.

Questions

1 $2Mg\,(s) + SO_2\,(g) \longrightarrow 2MgO\,(s) + S\,(s)$
In this reaction, which substance is:
a oxidized? **b** the oxidizing agent?
c reduced? **d** the reducing agent?

2 Copy and complete: Combustion usually means burning in .... The burning substance is ...... and one or more ...... are produced. ...... is also given out. Combustion in body cells is called ....

6.6 Different types of chemical reaction (III)

Cracking

Crude oil is a mixture of hundreds of different **hydrocarbons**. These are compounds made from carbon and hydrogen. They usually have long chains of carbon atoms, with hydrogen bonded on.

The diagram below shows a molecule of a typical hydrocarbon. When the compound is heated in the presence of a **catalyst** (a substance which speeds up a reaction), its molecules break down into smaller molecules. This process is called **cracking**.

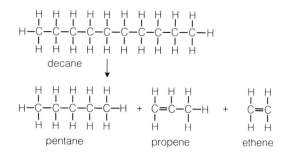

Note that two of the products have **double bonds** between carbon atoms. That makes them reactive. Cracking always gives products with double bonds. These are then turned into other products such as plastics, as you will see below. (See also page 194.)

Polymerization

One of the products from the reaction above is ethene, a colourless gas. Look what happens when it is heated at very high pressure:

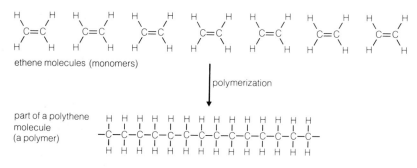

The double bonds break, and the ethene molecules join up to make very long molecules with thousands of carbon atoms. The result is **polythene**, the plastic used for plastic bags.
Polythene is a **polymer**. (*Poly-* means *many*.) The small starting molecules are called **monomers**. The reaction is called a **polymerization**.
Polymerization is used for making all kinds of plastics. On page 195 you can see how it is used for making PVC.

One of the many uses for polymers.

High-density and low-density plastics

Look what can happen if you change the reaction conditions for polymerizing ethene:

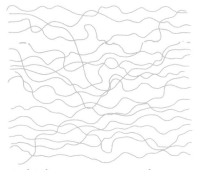

At high temperature and pressure, but without a catalyst, you get long chains like these. This is called **low-density** polythene.

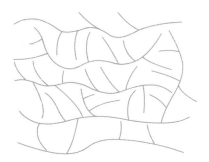

With a catalyst, and much lower temperature and pressure, you can make the chains branch. This is called **high-density** polythene.

So by choosing the right conditions you can change the density of the polythene, and make it 'heavy' or 'light' to suit your needs.

Addition and condensation polymerization

When ethene polymerizes to give polythene, the ethene molecules just add on to each other. So the polymerization is an **addition polymerization**.

Nylon is also a polymer. But this time two *different* monomers join:

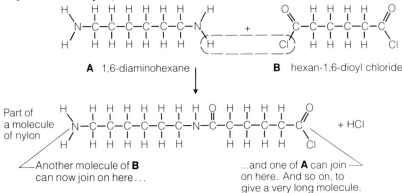

A 1,6-diaminohexane   **B** hexan-1,6-dioyl chloride

Part of a molecule of nylon

+ HCl

Another molecule of **B** can now join on here …

…and one of **A** can join on here. And so on, to give a very long molecule.

The monomers join up, but a molecule of hydrogen chloride is released at the same time. This is called a **condensation polymerization**.

If a small molecule is released during polymerization, the reaction is a condensation polymerization.

The nylon above is called 6,6-nylon. Can you see why? Try counting!

A rope made from nylon is flexible and very strong.

Questions

1 A straight-chain molecule, $C_{12}H_{26}$, was cracked to form three molecules. Two were C_2H_4 and C_3H_6.
   a Draw the starting molecule and the products. (Each carbon atom must have four bonds.)
   b What is the formula for the third product?

2 Draw a diagram to show how molecules of:
   a ethene, C_2H_4 **b** propene, C_3H_6
   could polymerize to make longer chains.

3 What is the difference between *addition polymerization* and *condensation polymerization*?

6.7 Energy changes in chemical reactions

Breaking and forming bonds

A mixture of hydrogen and chlorine will explode in sunshine.
The equation for the reaction is:

hydrogen   +   chlorine $\longrightarrow$ hydrogen chloride
$$H_2(g)\quad+\quad Cl_2(g)\quad\longrightarrow\quad 2HCl(g)$$

The explosion is a sign that a lot of energy is given out. But where does it come from? Let's look more closely at what happens:

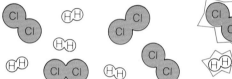 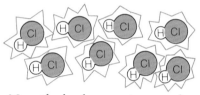

These are hydrogen and chlorine molecules. The atoms are held together by single bonds. These must be broken before reaction can take place. This needs energy.

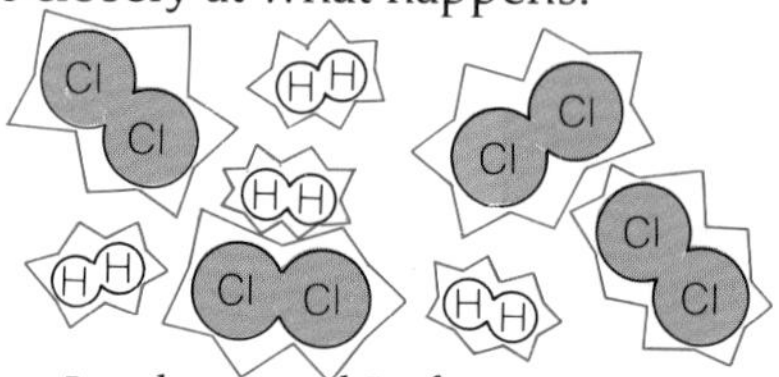

It takes 436 kJ of energy to break the bonds in one mole of hydrogen molecules and 242 kJ to break them in one mole of chlorine molecules. Energy from light starts the process off.

Now the hydrogen atoms and chlorine atoms bond. This *gives out* energy: 431 kJ when one mole of hydrogen chloride molecules forms, so 862 kJ for *two* moles.

Note that energy is measured in **kilojoules**, or **kJ** for short.
Now let's do our sums:

Energy in to break bonds:
    for a mole of hydrogen molecules 436 kJ
    for a mole of chlorine molecules <u>242 kJ</u>
    Total energy in 678 kJ

Energy out from forming bonds:
    for 2 moles of hydrogen chloride 862 kJ

Energy out − energy in   862 kJ
                     <u>− 678 kJ</u>
                     184 kJ

So the reaction gives out 184 kJ more energy than it takes in.
This is roughly the amount of energy given out by a one-bar electric fire if you run it for three minutes, but here it takes less than a second!
Since it gives out energy, the reaction is **exothermic**.

Bond energies

You saw above that it takes 242 kJ to break the bonds in a mole of chlorine molecules. This is called the **bond energy** of chlorine.
**Bond energy is the energy needed to break one mole of bonds.
It is measured in kilojoules per mole or kJ/mole.**

If these chlorine atoms were to bond to each other again, exactly the same amount of energy would be given out: 242 kJ.
The bond energy is also the energy *given out* when a mole of bonds is formed.
Some bond energies are given on the right.

Bond energies in kJ/mole	
C—C	346
C=C	612
C—O	358
O=O	498
O—H	464
H—H	436
N≡N	946
N—H	391

An endothermic reaction

If you blow steam through white-hot coke, the carbon gets oxidized:

carbon + water vapour $\longrightarrow$ carbon monoxide + hydrogen
$Cs(s)$ + $H_2O(g)$ $\longrightarrow$ $CO(g)$ + $H_2(g)$

The calculation on the right shows that this reaction *takes in* more energy than it gives out: 132 kJ more. So it is **endothermic**.

Energy in to break bonds:	
1 mole of carbon	717 kJ
1 mole of water	928 kJ
Total energy in	**1645 kJ**

Energy out from forming bonds:

1 mole of carbon	
monoxide	1077 kJ
1 mole of hydrogen	436 kJ
Total energy out	**1513 kJ**

| **Energy in − energy out =** | **132 kJ** |

Drawing an energy diagram

You can show the energy change in a reaction on an **energy diagram**.

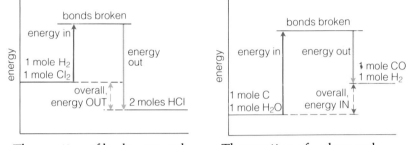

The reaction of hydrogen and chlorine. The products are at a lower energy level than the reactants. This shows that the reaction is exothermic.

The reaction of carbon and steam. The products are at a higher energy level than the reactants. So the reaction is endothermic.

Activation energy Even an exothermic reaction may need energy to start it off. This is called **activation energy**. It is the mimimum energy needed to break enough bonds to get the reaction started. It could come from sunlight, or a lit match, or a Bunsen burner.

The burning of fuels

Fuels give out energy when they burn − combustion is exothermic. Compare the products from burning these fuels. What do you notice?

Fuel	Used for	Equation for burning in oxygen	Energy (kJ) from burning: 1 mole of fuel	1 gram of fuel
Methane	heating, cooking	$CH_4(g) + 2O_2(g) \longrightarrow CO_2(g) + 2H_2O(g)$	890	56
Butane	camping gas	$2C_4H_{10}(g) + 13O_2(g) \longrightarrow 8CO_2(g) + 10H_2O(g)$	2877	50
Hydrogen	rocket fuel	$2H_2(g) + O_2(g) \longrightarrow 2H_2O(g)$	242	121
Glucose	your body fuel	$C_6H_{12}O_6(aq) + 6O_2(g) \longrightarrow 6CO_2(g) + 6H_2O(g)$	2803	16

Now compare the energies given out when they burn. Why do you think hydrogen is chosen as a rocket fuel instead of glucose?

Questions

1 Explain what happens to the bonds when hydrogen reacts with chlorine. Use drawings to help you.
2 Why does this reaction *give out* energy?

3 Draw an energy diagram for the combustion of North Sea gas in a gas cooker. How would you provide the activation energy for the reaction?

Questions on Section 6

1 Decide whether each change below is a physical change or a chemical change. Give reasons for your answers:
   a Ice melting.
   b Iron rusting.
   c Petrol burning.
   d Candle wax melting.
   e A candle burning.
   f Wet hair drying.
   g Milk souring.
   h Perfume evaporating.
   i A lump of roll sulphur being crushed.
   j Copper being obtained from copper(II) oxide.
   k Clothes being ironed.
   l Custard being made.
   m A cigarette being smoked.

2 Write equations for the following reactions:
   a 1 mole of copper atoms combines with 1 mole of sulphur atoms to form 1 mole of copper(II) sulphide, CuS.
   b 3 moles of lead atoms combines with 2 moles of oxygen molecules to form 1 mole of lead oxide, Pb_3O_4.
   c 1 mole of ethanol molecules, C_2H_5OH, burns in 3 moles of oxygen molecules to form 2 moles of carbon dioxide molecules and 3 moles of water molecules.
   d 1 mole of iron(III) oxide, Fe_2O_3, is reduced by 3 moles of hydrogen molecules to form 2 moles of iron atoms and 3 moles of water molecules.

3 Balance the following equations.
   a Synthesis of hydrogen bromide:
   $$H_2(g) + Br_2(g) \longrightarrow HBr(g)$$
   b Decomposition of lead nitrate:
   $$Pb(NO_3)_2(s) \xrightarrow{heat} PbO(s) + NO_2(g) + O_2(g)$$
   c Electrolysis of aluminium oxide:
   $$Al_2O_3(l) \longrightarrow Al(l) + O_2(g)$$
   d Precipitation of barium sulphate:
   $$BaCl_2(aq) + Na_2SO_4(aq) \longrightarrow BaSO_4(s) + NaCl(aq)$$
   e Neutralization of sodium hydroxide:
   $$NaOH(aq) + H_2SO_4(aq) \longrightarrow Na_2SO_4(aq) + H_2O(l)$$
   f Reduction of carbon dioxide:
   $$Mg(s) + CO_2(g) \longrightarrow MgO(s) + C(s)$$
   g Combustion of carbon monoxide:
   $$CO(g) + O_2(g) \longrightarrow CO_2(g)$$
   h Combustion of methane (natural gas):
   $$CH_4(g) + O_2(g) \longrightarrow CO_2(g) + H_2O(g)$$
   i Cracking of ethane:
   $$C_2H_6(s) \longrightarrow C_2H_4(s) + H_2(s)$$
   j Displacement of silver:
   $$Zn(s) + AgNO_3(aq) \longrightarrow Zn(NO_3)_2(aq) + Ag(s)$$

4 Mercury(II) oxide breaks down into mercury and oxygen when heated. The equation for the reaction is:
   $$2HgO(s) \longrightarrow 2Hg(l) + O_2(g)$$
   a Calculate the mass of 1 mole of mercury(II) oxide. ($O = 16$, $Hg = 201$).
   b Find out how much mercury and oxygen are produced when 21.7 g of mercury(II) oxide is heated.

5 Iron(II) sulphide is formed when iron and sulphur react together:
   $$Fe(s) + S(s) \longrightarrow FeS(s)$$
   a How many grams of sulphur will react with 56 g of iron? (The RAMs on page 73 will help you.)
   b If 7 g of iron and 10 g of sulphur are used, which substance is in excess?
   c If 7 g of iron and 10 g of sulphur are used, name the substances present when the reaction is complete, and find the mass of each.
   d Is the reaction endothermic or exothermic? Explain.

6 The following equation represents a reaction in which iron is obtained from iron(III) oxide:
   $$Fe_2O_3(s) + 3CO(g) \xrightarrow{heat} 2Fe(s) + 3CO_2(g)$$
   a Which substance is reduced?
   b Which substance is oxidized?
   c What is the formula mass of iron(III) oxide? ($Fe = 56$, $O = 16$)
   d How many moles of Fe_2O_3 are present in 320 kg of iron(III) oxide? (1 kg = 1000 g)
   e How many moles of Fe are obtained from 1 mole of Fe_2O_3?
   f From d and e, find the number of moles of iron obtained from 320 kg of iron(III) oxide.
   g Find the mass of iron obtained from 320 kg of iron(III) oxide.

7 When solutions of potasssium sulphate and barium chloride were mixed, a white precipitate formed. The equation for the reaction is:
   $$K_2SO_4(aq) + BaCl_2(aq) \longrightarrow BaSO_4(s) + 2KCl(aq)$$
   22.3 g of barium sulphate was obtained.
   a What is the formula mass of barium sulphate? ($Ba = 137$, $S = 32$, $O = 16$)
   b How many moles of $BaSO_4$ is 23.3 g of barium sulphate?
   c How many moles of K_2SO_4 and $BaCl_2$ must have reacted, to form 23.3 g of barium sulphate?
   d The concentrations of the two reacting solutions were 0.1 M (that is, 0.1 moles of solute in 1 litre of solution). What volume of each solution was needed to form 23.3 g of barium sulphate?
   e How could the precipitate be removed?

For questions 8 to 11 you will need to remember that:
The volume of one mole of any gas measured at room temperature and pressure (rtp) is 24 dm³ or 24 000 cm³.

8 What is the volume at rtp in dm³ and cm³, of:
a 2 moles of hydrogen, H_2?
b 0.5 moles of carbon dioxide, CO_2?
c 0.1 moles of nitrogen, N_2?
d 0.3 moles of oxygen, O_2?

9 Baking powder contains sodium hydrogen carbonate, $NaHCO_3$. When it is heated it decomposes as follows:
$$2NaHCO_3\,(s) \longrightarrow Na_2CO_3\,(s) + H_2O\,(l) + CO_2\,(g)$$
a Write a word equation for the reaction.
b i How many moles of sodium hydrogen carbonate are on the left side of the equation?
ii What is the mass of this amount? ($Na = 23, H = 1, C = 12, O = 16$)
c i How many moles of carbon dioxide are on the right side of the equation?
ii What is the volume of this amount at rtp?
d What volume at rtp of carbon dioxide would be obtained if:
i 84 g of sodium hydrogen carbonate was completely decomposed?
ii 8.4 g of sodium hydrogen carbonate was completely decomposed?
e Explain how baking powder helps cakes rise during baking.

10 Limestone is calcium carbonate, $CaCO_3$. When limestone is heated, this chemical change occurs:
$$CaCO_3\,(s) \longrightarrow CaO\,(s) + CO_2\,(g)$$
($Ca = 40, C = 12, O = 16$)
a Write a word equation for the chemical change.
b How many moles of $CaCO_3$ are there in 50 g of calcium carbonate?
c i What mass of calcium oxide is obtained from the thermal decomposition of 50 g of calcium carbonate?
ii What mass of carbon dioxide would be given off at the same time?
iii What volume would the gas occupy at rtp?

11 Hydrogen peroxide is a colourless liquid with strong oxidizing properties, used as a bleach. It decomposes like this:
$$2H_2O_2\,(aq) \longrightarrow 2H_2O\,(l) + O_2\,(g)$$
It is sold as a 3% solution, which means that 1 litre of solution contains 30 g of hydrogen peroxide.
a Why do bubbles appear in the solution as the hydrogen peroxide decomposes?
b i What is the mass of 1 mole of hydrogen peroxide? ($H = 1, O = 16$)
ii What is the concentration in moles per litre of a 3% solution?
iii How many moles of oxygen molecules are formed when 1 litre of the solution decomposes?
iv What volume of oxygen is this at rtp?

12 Calor gas is a hydrocarbon called propane, C_3H_8. In an experiment, 1 gram of propane was used to heat 1000 cm³ of water. The temperature of the water rose from 20 °C to 32 °C.
a By how much did the water temperature rise?
b Is the burning of Calor gas exothermic or endothermic?
c i Calculate the heat energy given out when 1 gram of propane burns. (4.2 kJ raises the temperature of 1000 cm³ of water by 1 °C)
ii What is the heat energy given out when 1 mole of propane burns? ($C = 12, H = 1$)
Lighter fuel is another hydrocarbon, called butane (C_4H_{10}). When 1 mole of butane burns, 2900 kJ of heat energy is given out.
d Use the idea of making and breaking bonds to explain why burning 1 mole of butane produces more energy than burning 1 mole of propane.

13 When ammonium chloride (NH_4Cl) dissolves in water, the temperature drops by a few degrees.
a i Is the dissolving exothermic or endothermic?
ii What can you say about the bonds that are formed in the solution?
When 1 mole of ammonium chloride dissolves in water, 15 kJ of energy is taken in.
b How much energy is taken in when: **i** 107 g
ii 1 kg of ammonium chloride dissolves in water?

14 Methane is the main component of natural gas. The reaction between methane and oxygen is exothermic:
$$CH_4\,(s) + O_2\,(g) \longrightarrow CO_2\,(g) + 2H_2O\,(g)$$
a Explain in terms of bond breaking and bond making why this reaction is exothermic.
b i Copy the diagram below and complete it to show the energy diagram for this reaction.

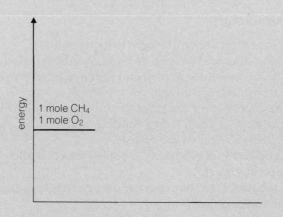

ii Methane will not burn in air until a spark or flame is applied. Why not?
When 1 mole of methane burns in oxygen, 890 kJ of energy is given out.
c How much energy is given out when 1 g of methane burns? ($C = 12, H = 1$)

7.1 Conductors and insulators

Batteries and electric current

The photograph below shows a battery, a bulb and a rod of graphite (carbon) joined or **connected** to each other by copper wires. The arrangement is called an **electric circuit**. The bulb is lit: this shows that electricity must be flowing in the circuit.
Electricity is a stream of moving electrons.
Do you remember what an electron is? It is a tiny particle with a negative charge and almost no mass.

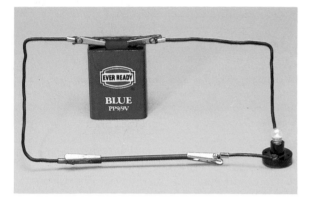

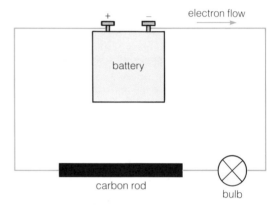

The diagram above right shows how the electrons move through the circuit. The battery acts like an electron pump. Electrons leave it through one terminal, called the **negative terminal**. They are pumped through the wires, the bulb and the rod, and enter the battery again through the **positive terminal**. When the electrons stream through the fine wire in the bulb, they cause the wire to heat up so much that it gets white-hot and gives out light.

Conductors

In the circuit above, the light will go out if:
● you disconnect a wire, so that everything no longer joins up. *Or*
● you connect something into the circuit that prevents electricity from flowing through.
The copper wires and graphite rod obviously do allow electricity to flow through them – they **conduct** electricity. Copper and graphite are therefore called **conductors**.
A conductor of electricity is a substance that allows electricity to flow through it.
A substance that does not conduct electricity is called a **non-conductor** or **insulator**.

Testing substances to see if they conduct

The circuit above can be used to test any substance to see if it conducts electricity. The substance is simply connected into the circuit, like the graphite rod above. Some examples are given on the next page.

Copper is the conductor inside this electric drill. But plastic (an insulator) is used to connect the bit to the motor, and for the outer case. Why is that?

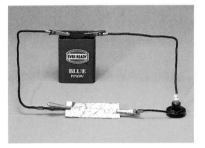

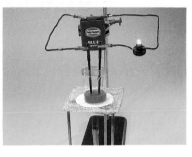

Testing tin to see if it conducts electricity. A strip of tin is connected into the circuit. The bulb lights, so tin must be a conductor.

Testing ethanol. The liquid is connected into the circuit by dipping graphite rods into it. The bulb does not light, so ethanol is a non-conductor.

Testing molten lead bromide. A Bunsen is used to melt it. The molten compound conducts, and at the same time gives off a choking brown vapour.

The results These are the results from a range of tests:

1 **The only solids that conduct are the metals and graphite.**
   These conduct because of their free electrons (pages 56 and 57). The free electrons get pumped out of one end of the solid by the battery. Electrons then flow in the other end, and through the spaces left behind.
   For the same reason, *molten* metals conduct. (It is not possible to test molten graphite, because graphite sublimes.)

2 **Molecular or covalent substances are non-conductors.**
   This is because they contain no free electrons, or other charged particles, that can flow out through them.
   The ethanol above is one example of a molecular substance. Others are petrol, paraffin, sulphur, sugar and plastic. These never conduct, whether solid or molten.

3 **Ionic substances do not conduct when solid. However, they conduct when melted or dissolved in water, and they decompose at the same time.**
   An ionic substance contains no free electrons. However, it does contain ions, which are also charged particles. The ions become free to move when the substance is melted or dissolved, and it is they that conduct the electricity.
   The lead bromide above is an example. It is a non-conductor when solid. But it begins to conduct the moment it is melted, and a brown vapour bubbles off at the same time. The vapour is bromine, and it forms because electricity causes the lead bromide to **decompose**.
   Decomposition caused by electricity is called electrolysis, and the liquid that decomposes is called an electrolyte.
   Molten lead bromide is therefore an electrolyte. Ethanol is a **non-electrolyte** because it does not conduct at all.

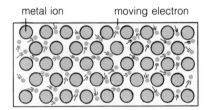

metal ion   moving electron

Like other metals, aluminium conducts electricity. It is used for electricity cables because it is so light.

Questions

1 What is a **conductor** of electricity?
2 Draw a circuit to show how you would test whether mercury conducts.
3 Explain why metals are able to conduct electricity.

4 Naphthalene is a molecular substance. Do you think it conducts when molten? Explain why.
5 What is: **a** an electrolyte? **b** a non-electrolyte? Give *three* examples of each.

7.2 A closer look at electrolysis

The electrolysis of lead bromide

On the last page, you saw that molten lead bromide decomposes when it conducts electricity. Decomposition caused by electricity is called **electrolysis**, and molten lead bromide is an **electrolyte**.

The apparatus   This is shown on the right. The graphite rods carry the current into and out of the molten lead bromide. Conducting rods like these are called **electrodes**. The electrode joined to the negative terminal of the battery is called the **cathode**. It is also negative, because the electrons from the battery flow to it. The other electrode is positive, and is called the **anode**. Notice the switch. When it is open, no electricity can flow.

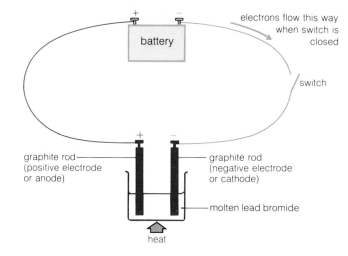

The electrolysis   Once the switch is closed, **bromine vapour** starts to bubble out of the molten lead bromide, around the anode. After some time a bead of **molten lead** forms below the cathode. The electrical energy from the battery has caused a **chemical change**:

lead bromide $\longrightarrow$ lead $+$ bromine
$$PbBr_2 (l) \longrightarrow Pb (l) + Br_2 (g)$$

Why the molten lead bromide decomposes   When lead bromide melts, its lead ions (Pb^{2+}) and bromide ions (Br^-) become free to move about. When the switch is closed, the electrodes become charged, and the ions are immediately attracted to them:

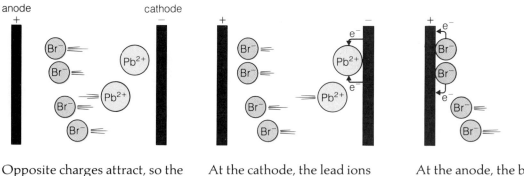

Opposite charges attract, so the lead ions are attracted to the cathode, and the bromide ions to the anode.

At the cathode, the lead ions each receive 2 electrons and become lead atoms:
$$Pb^{2+} + 2e^- \longrightarrow Pb$$
The lead atoms collect together on the cathode, and in time fall to the bottom of the beaker.

At the anode, the bromide ions each give up 1 electron to become bromine atoms. These pair together as molecules:
$$2\,Br^- \longrightarrow Br_2 + 2e^-$$
The bromine bubbles off as a gas.

Why the molten lead bromide conducts During the electrolysis, each lead ion takes two electrons from the cathode, as shown on the right. At the same time two bromide ions each give an electron to the anode. The effect is the same as if two electrons *flowed through the liquid* from the cathode to the anode. In other words, the lead bromide is acting as a conductor of electricity.

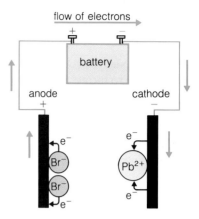

The electrolysis of other compounds

All ionic compounds can be electrolysed, when they are molten. (Another word for molten is **fused**.) These are some points to remember about the process:

1 The electrolyte always **decomposes**. So electrical energy is causing a **chemical change**.
2 The electrodes are usually made of graphite or platinum. These substances are unreactive or **inert**. That means they will not react with the electrolyte or the products of the electrolysis.
3 Metals always form positive ions. Positive ions always go to the cathode. So they are also called **cations**.
4 Nonmetals (except hydrogen) always form negative ions. They always go to the anode so are called **anions**.
5 **So when a molten ionic compound is electrolysed, a metal is always formed at the cathode and a nonmetal at the anode.**
You can see more examples in this table:

Electrolyte	The decomposition		At the cathode	At the anode
Sodium chloride NaCl	sodium chloride $\longrightarrow$ $2NaCl\,(l)$ $\longrightarrow$	sodium + chlorine $2Na\,(l) +\ Cl_2\,(g)$	$2Na^+ + 2e^- \longrightarrow 2Na$	$2Cl^- \longrightarrow Cl_2 + 2e^-$
Potassium iodide KI	potassium iodide $\longrightarrow$ $2KI\,(l)$ $\longrightarrow$	potassium + iodide $2K\,(l)\ \ +I_2\,(g)$	$2K^+ + 2e^- \longrightarrow 2K$	$2I^- \longrightarrow I_2 + 2e^-$
Copper(II) bromide $CuBr_2$	copper(II) bromide $\longrightarrow$ $CuBr_2\,(l)$ $\longrightarrow$	copper + bromine $Cu\,(l) + Br_2\,(g)$	$Cu^{2+} + 2e^- \longrightarrow Cu$	$2Br^- \longrightarrow Br_2 + 2e^-$

6 Electrolysis is the most powerful way to decompose an ionic compound. So it is used in industry to extract metals such as sodium and aluminium from their ores. (Sodium is extracted from molten rock salt or sodium chloride. Aluminium is extracted from molten aluminium oxide.) But electrolysis guzzles up electricity, which makes it very expensive. So it is used only for very stable compounds which are difficult to decompose in any other way. You can find out more on page 152.

Questions

1 Explain what each of these words means:
   electrolysis   anode   cathode
2 For the electrolysis of molten lead bromide, draw diagrams to show:
   a how the ions move when the switch is closed;
   b what happens at the anode;
   c what happens at the cathode.

3 What is: a cation?   an anion?
4 Molten copper and molten copper(II) bromide both conduct. What changes would you expect to see when they conduct?
5 Write equations for the overall reaction, and the reaction at each electrode, when fused magnesium chloride ($MgCl_2$) is electrolysed.

7.3 The electrolysis of solutions

When a salt such as sodium chloride is dissolved in water, its ions become free to move. So the solution can be electrolysed. But the products may be different from when you electrolyse the *molten* salt, *because water itself also produces ions*. Although water is molecular, a tiny fraction of its molecules is split into ions:

some water molecules $\longrightarrow$ hydrogen ions + hydroxide ions

$$H_2O\,(l) \longrightarrow H^+(aq) + OH^-(aq)$$

During electrolysis, these H^+ and OH^- ions compete with the metal and nonmetal ions from the dissolved salt, to receive or give up electrons. So which ions win? These are the rules.

At the cathode:
1 The more reactive a metal, the more it 'likes' to exist as ions. So if a metal is very reactive, its ions remain in solution. The H^+ ions accept electrons, and hydrogen molecules are formed.
2 The ions of less reactive metals will accept electrons and form metal atoms, leaving the H^+ ions in solution.

At the anode:
3 If ions of a halogen are present (Cl^-, Br^- or I^-), they will give up electrons more readily than the OH^- ions do. Molecules of chlorine, bromine or iodine are formed.
4 If no halogen ions are present, OH^- ions will give up electrons more readily than other nonmetal ions do, and oxygen is formed.

Now let's look at some examples.

Use apparatus like this when you want to collect the gases for electrolysis.

Sodium chloride solution

A concentrated solution of sodium chloride is electrolysed in the lab using the equipment shown above right. This is what happens:

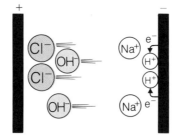
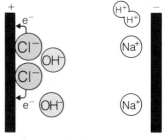

The solution contains Na^+ ions and Cl^- ions from the salt, and H^+ and OH^- ions from water. The positive ions go to the cathode and the negative ions to the anode.

At the cathode, it is the H^+ ions which accept electrons since sodium is more reactive than hydrogen:

$$2H^+ + 2e^- \longrightarrow H_2$$

Hydrogen gas bubbles off while Na^+ ions remain in solution.

At the anode, the Cl^- ions give up electrons more readily than the OH^- ions do. Chlorine gas bubbles off:

$$2Cl^- \longrightarrow Cl_2 + 2e^-$$

The OH^- ions remain in solution.

When the hydrogen and chlorine bubble off, Na^+ and OH^- ions are left behind: a solution of sodium hydroxide is formed.

Other salt solutions

Here are the results of electrolysing some other salt solutions.
Check them out. Do they obey the rules given opposite?

	When the electrolyte is a solution of . . .	At the cathode you get . . .	At the anode you get . . .
	potassium bromide KBr (aq)	hydrogen $2H^+ + 2e^- \longrightarrow H_2$	bromine $2Br^- \longrightarrow Br_2 + 2e^-$
	sodium iodide NaI (aq)	hydrogen $2H^+ + 2e^- \longrightarrow H_2$	iodine $2I^- \longrightarrow I_2 + 2e^-$
metals increasingly reactive	magnesium sulphate $MgSO_4$ (aq)	hydrogen $4H^+ + 4e^- \longrightarrow 2H_2$	oxygen $4OH^- \longrightarrow 2H_2O + O_2 + 4e^-$
	lead(II) nitrate $Pb(NO_3)_2$ (aq)	lead $2Pb^{2+} + 4e^- \longrightarrow 2Pb$	oxygen $4OH^- \longrightarrow 2H_2O + O_2 + 4e^-$
	copper(II) chloride $CuCl_2$ (aq)	copper $Cu^{2+} + 2e^- \longrightarrow 2Cu$	chlorine $2Cl^- \longrightarrow Cl_2 + 2e^-$
	silver nitrate $AgNO_3$ (aq)	silver $4Ag^+ + 4e^- \longrightarrow 4Ag$	oxygen $4OH^- \longrightarrow 2H_2O + O_2 + 4e^-$

Dilute sulphuric acid

Sulphuric acid has the formula H_2SO_4. In water it forms ions:
$$H_2SO_4 (aq) \longrightarrow 2H^+ (aq) + SO_4{}^{2-} (aq)$$
As you've seen, water also produces ions. So a dilute solution of the
acid contains H^+ ions from both water and acid, OH^- ions from the
water and $SO_4{}^{2-}$ ions from the acid. It can be electrolysed using
apparatus like that shown on the right or on the opposite page.

At the cathode This time there are no metal ions to compete
with. So hydrogen gas is formed and bubbles off:
$$4H^+ + 4e^- \longrightarrow 2H_2$$

At the anode The OH^- and $SO_4{}^{2-}$ ions compete to give up
electrons. As you'd expect from rule 4, the OH^- ions win:
$$4OH^- \longrightarrow 2H_2O + O_2 + 4e^-$$
The oxygen bubbles off. The $SO_4{}^{2-}$ ions are left behind in solution.

The overall result is that the water decomposes rather than the acid:
water $\longrightarrow$ hydrogen + oxygen
$$2H_2O (l) \longrightarrow 2H_2 (g) + O_2 (g)$$
So this electrolysis is often called **the electrolysis of acidified water**.

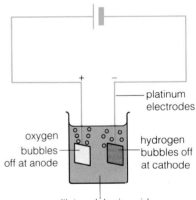

platinum electrodes

oxygen bubbles off at anode

hydrogen bubbles off at cathode

dilute sulphuric acid

Use apparatus like this when you don't
want to collect the gases from
electrolysis.

Questions

1 Explain why water conducts electricity very slightly.
2 Describe what happens during the electrolysis of a
  concentrated solution of sodium chloride.
3 Which ions are there in a solution of sodium iodide?

4 Write an equation for the reaction you expect at each
  electrode, when you electrolyse a solution of:
  a sodium nitrate, $NaNO_3$ **b** hydrochloric acid, HCl
  c potassium iodide, KI **d** copper sulphate, $CuSO_4$

7.4 Using electrolysis to deposit metals

Copper(II) sulphate solution and copper electrodes

When a solution of copper(II) sulphate is electrolysed using carbon or platinum electrodes, copper is obtained at the cathode and oxygen at the anode, as you'd expect from the rules on page 102. But when *copper* electrodes are used instead, something unusual happens.

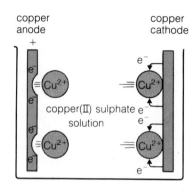

At the cathode   Copper ions become atoms, as you would expect:
$$Cu^{2+} + 2e^- \longrightarrow Cu$$
The copper atoms cling to the cathode.

At the anode   The copper anode *dissolves*, forming copper ions:
$$Cu \longrightarrow Cu^{2+} + 2e^-$$
So the anode wears away, while the cathode grows thicker. If you weigh them you will find that:

mass of copper lost by anode = mass of copper gained by cathode

This electrolysis is used in industry to purify copper, which has to be very pure for use in electric wires. The impure copper is made into an anode. As it wears away, the cathode gets plated with pure copper while impurities fall to the bottom of the electrolysis cell.

What affects the amount of copper deposited?

During the electrolysis, copper atoms give up electrons to the anode. The battery acts as an electron pusher, pushing these electrons to the cathode as an electric current. There copper ions take them and form a deposit of copper atoms. So:

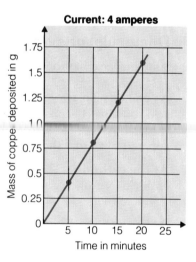

1 **The longer the current flows, the more copper is deposited.**
   This is shown on the first graph on the right. If you double the time, you will double the mass deposited.
2 **The larger the current, the more copper is deposited.**
   This is shown on the second graph on the right. Doubling the current for a given time will double the mass deposited.

So the amount of copper deposited depends on the size of the current and how long it runs.

A closer look at current

A current is a flow of electrons. It is measured in units called **amperes** (**A**). Each electron has a very tiny electric charge. The unit of measurement for the total charge of the current is the **coulomb (C)**. **A current of 1 ampere flowing for 1 second carries a charge of 1 coulomb.**

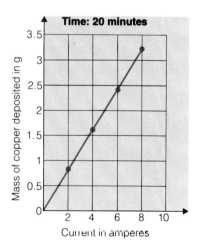

To find the charge of any current we can use this formula:
   current (A) × time (s) = charge in coulombs (C)

How is the charge on an electron related to the coulomb? Like this:
1 mole of electrons has a charge of 96 500 coulombs.

Calculating the mass of copper deposited

A solution of copper(II) sulphate is electrolysed using copper electrodes. How much copper will be deposited by a current of 2 A flowing for 20 minutes?

Step 1 Calculate the charge of the current.

Charge (coulombs) = time (seconds) × current (amperes)
= 20 × 60 × 2 = 2400 coulombs

Step 2 Convert the charge to moles of electrons.

96 500 coulombs = 1 mole of electrons

so 1 coulomb $= \dfrac{1}{96\,500}$ moles of electrons

2400 coulombs $= \dfrac{1}{96\,500} \times 2400$ moles of electrons

= 0.025 moles of electrons

Step 3 Now use the equation for the reaction at the cathode to find how much copper is deposited.

$Cu^{2+} + 2e^- \longrightarrow Cu$

So 2 moles of electrons gives 1 mole of copper atoms.
Changing moles of copper atoms to grams:
2 moles of electrons gives 64 g of copper so
1 mole of electrons gives 32 g and
0.025 moles of electrons gives (0.025 × 32) g or 0.8 g.
So 0.8 grams of copper will be deposited.

The apparatus for depositing copper. Which is the positive electrode? Which will get heavier?

Electroplating

You can also use electrolysis to coat one metal with a *different* one. This is called **electroplating**.
Suppose you want to coat a nickel jug with silver. You could set up an electrolysis cell with the jug as the cathode and silver foil as the anode, in a solution of silver nitrate.

At the anode   The silver dissolves, forming silver ions:
$Ag \longrightarrow Ag^+ + e^-$

At the cathode   Silver ions receive electrons and form a layer of silver on the jug. When the layer is thick enough the jug is removed:
$Ag^+ + e^- \longrightarrow Ag$

Electroplating is used a great deal in industry. For example, car bumpers are plated with chromium to make them look good and to protect them from rust. Steel is plated with tin to make 'tins' for food.

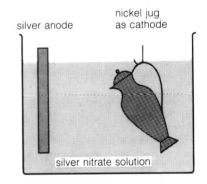

silver anode

nickel jug as cathode

silver nitrate solution

THE RULES
To electroplate an object with metal X, use:
● the object as cathode
● a strip of X as anode
● a solution of a compound of X as electrolyte

Questions

1 Copper(II) sulphate solution is electrolysed for 15 minutes using copper electrodes and a current of 4 A.
   a What charge is carried by the current?
   b How many moles of electrons does this equal?
   c How much copper is deposited?
   d The solution remains blue. Why is this?

2 Draw a diagram to show how you would purify copper using electrolysis.
3 a Draw a labelled diagram to show how you would plate an iron nail with nickel.
   b You could use nickel chloride for the electrolyte, but not nickel carbonate. Why is this?

7.5 The chlor-alkali industry (I)

A whole section of the chemical industry has developed around a single electrolysis reaction. It's called the **chlor-alkali** industry, and it is based on the electrolysis of a solution of sodium chloride, or **salt**.

Most of the salt in Britain comes from Cheshire, where it lies 200 metres below ground in seams 200 m thick. Some is mined as rock salt for gritting roads. But most is pumped out of the ground in a concentrated solution called **brine**, which is then electrolysed:

$$\text{brine} \longrightarrow \text{sodium hydroxide} + \text{chlorine} + \text{hydrogen}$$
$$2NaCl\,(aq) + 2H_2O\,(l) \longrightarrow 2NaOH + Cl_2\,(g) + H_2\,(g)$$

The products of the electrolysis are used in turn as raw materials for a whole range of other products.

Brine – the starting point for many new materials.

Getting the salt out from underground

First, three concentric pipes are buried 400 metres into the ground. Water is pumped down the outer pipe to dissolve the salt. Air is pumped down the second. The brine is forced up the inner pipe and into reservoirs, then piped to a purification plant when it is needed.

The brine is purified by adding sodium hydroxide to precipitate metal ions as hydroxides. (Most hydroxides are insoluble.) Sulphate ions are removed by adding barium ions to precipitate barium sulphate.

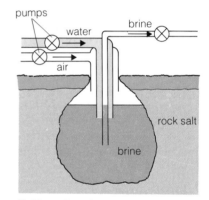

Getting salt out from underground.

Electrolysing the brine

The electrolysis is just like the one you can do in the lab (page 102), but on a much larger scale. The main thing is to make sure the hydrogen and chlorine are kept apart.

Different electrolysis cells can be used, but the **membrane cell** uses least electricity and is safest for the environment. The membrane cell has a titanium anode and a nickel cathode. It has an **ion-exchange membrane** up the middle. This lets sodium ions through but keeps the gases apart. So the sodium ions can move freely to the cathode.

At the cathode Hydrogen bubbles off:
$$2H^+ + 2e^- \longrightarrow H_2$$

At the anode Chlorine bubbles off:
$$2Cl^- \longrightarrow Cl_2 + 2e^-$$

Na$^+$ and OH$^-$ ions are left behind, which means a solution of sodium hydroxide forms. Some is evaporated to a more concentrated solution, and some evaporated completely to give solid sodium hydroxide.

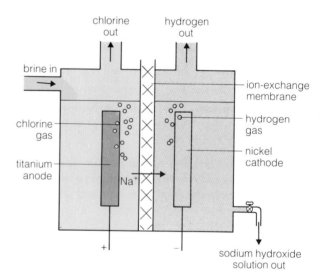

Chlorine is used as an anti-bacterial agent.

Sodium hydroxide is extremely reactive and must be handled with care.

What the products are used for

A concentrated solution of sodium chloride is a very simple starting material. You could make it in your kitchen. But look what it leads to!

Electrolysis of brine

Chlorine, a poisonous yellow-green gas

Used for making ...
- PVC (nearly $\frac{1}{3}$ of it used for this)
- solvents such as trichloroethane for degreasing and dry-cleaning
- paints and dyestuffs
- bleaches, weedkillers, pesticides
- pharmaceuticals
- titanium dioxide, a white pigment used in paints, ceramics, cosmetics and paper
- hydrogen chloride and hydrochloric acid

It is also used for ...
- killing bacteria in tap water
- killing bacteria in swimming pools

Hydrogen, a colourless flammable gas

Used for ...
- making nylon
- making hydrogen peroxide
- 'hardening' vegetable oils for margarines
- burning to make steam for other processes

Sodium hydroxide solution, alkaline and corrosive

Used for making ...
- soaps
- detergents
- viscose (rayon) and other textiles
- paper (such as used in this book)
and in making many chemicals including dyestuffs and pharmaceuticals

Questions

1 Write a word equation for the electrolysis of brine.
2 Draw a diagram for the membrane cell used for it.
3 Why is the membrane needed?
4 List four uses for: **a** sodium hydroxide **b** brine

5 Where does the term 'chlor-alkali' come from?
6 Suppose your job is to keep a brine electrolysis plant running safely and smoothly. Try to think of four safety precautions you might need to take.

7.6 The chlor-alkali industry (II)

A little history

Around two hundred years ago, Cheshire already had a lively chemical industry based on salt. In factories around Widnes, salt was turned into sodium carbonate. This was sold to Merseyside's soap-makers for making soap, and to its textile factories for treating textiles, and to the glass factories of St. Helens for making glass.

Then, in 1897, the first factory for electrolysing brine was opened at Runcorn. Now the soap-makers were able to buy sodium hydroxide instead of sodium carbonate for making their soap. The chlorine produced was sold for bleaching textiles and paper, and for making bleaching powder.

There was bitter rivalry between the different companies making chemicals from salt, and fierce competition to find new markets. But in 1926, the two largest joined up with a dye company and an explosives company, to form the giant Imperial Chemical Industries, or ICI.

Today, ICI at Runcorn still runs one of the world's largest plants for electrolysing brine.

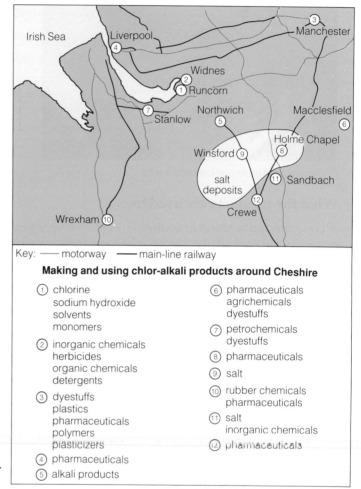

Key: —— motorway —— main-line railway

Making and using chlor-alkali products around Cheshire

① chlorine
    sodium hydroxide
    solvents
    monomers

② inorganic chemicals
    herbicides
    organic chemicals
    detergents

③ dyestuffs
    plastics
    pharmaceuticals
    polymers
    plasticizers

④ pharmaceuticals

⑤ alkali products

⑥ pharmaceuticals
    agrichemicals
    dyestuffs

⑦ petrochemicals
    dyestuffs

⑧ pharmaceuticals

⑨ salt

⑩ rubber chemicals
    pharmaceuticals

⑪ salt
    inorganic chemicals

⑫ pharmaceuticals

Is it a profitable business?

This shows what ICI make from a tonne (1000 kg) of salt:

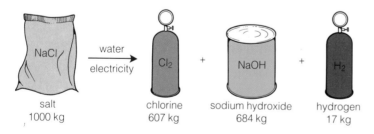

salt 1000 kg →(water, electricity)→ chlorine 607 kg + sodium hydroxide 684 kg + hydrogen 17 kg

Every time they make 607 kg of chlorine they also make 684 kg of sodium hydroxide — whether they can sell it or not!

The aim of this business, like any other, is to make a profit:
operating profit (£) = sales (£) − direct costs (£) − overheads (£)

So is there plenty of profit to be made from the electrolysis of brine? The answer is: it varies! The higher the sales and the lower the direct costs and overheads, the higher the profit will be.

ICI's Runcorn works provides much local employment and prosperity.

Direct costs

These are the costs of producing the products:

1 **Raw material**. The salt and water are readily available in Runcorn and quite cheap.

2 **Labour**. There's plenty of labour available too. But the production processes are automated, so don't need many people.

3 **Fuel**. The electrolysis goes on 24 hours a day, 365 days a year. Electricity is also used for pumping out the brine. So electricity is by far the biggest cost. ICI paid over £60 million for it in 1993.
If the bill keeps rising, it will have a big effect on profits. It will force the company to charge more for its products, and that in turn could force customers to find cheaper producers abroad.

4 **Research and development**. A team of scientists is continually working on ways to make the electrolysis plant more efficient, and ensure it does the environment no harm.

Overheads These are the general costs of running any business: the salaries of the managers, secretaries and sales team, the cost of running the canteen, the central heating, carpets, desks and so on. Good management will keep these costs well under control.

Sales The electrolysis produces both sodium hydroxide and chlorine, but these do not sell equally well. Sales depend on the economy. In a strong economy, more chlorine is needed because people buy more PVC goods, solvents and so on. Its price is pushed up because customers compete for it. But meanwhile the stocks of unsold sodium hydroxide grow, so its price is cut to encourage customers. The opposite happens in a recession.

One thing which will damage sales of chlorine in the future is concern about the environment. Chlorine escapes into the atmosphere in compounds such as **chlorofluorocarbons** (CFCs) and **hydrofluorocarbons** (HCFCs) which are used in aerosols, packaging foams, fridges, air-conditioning units, and as solvents. In the atmosphere, it triggers a chain reaction that destroys the ozone layer.
For this reason many chlorine compounds have already been banned, and others probably will be.

Membrane electrolysis technology.

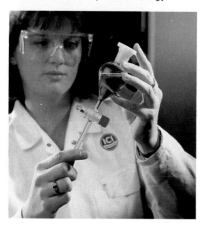

Research and development is an important aspect of the chlor-alkali industry.

Questions

1 Write down at least five reasons why Runcorn is a good location for the electrolysis of brine.

2 What do you think your concerns would be, if you:
 a lived close to a brine electrolysis plant?
 b had shares in the plant?
 c ran a factory needing large amounts of chlorine?

3 Suppose it is your job to manage the distribution of sodium hydroxide, chlorine and hydrogen from the plant to customers. What precautions would you take?

4 Some cans of hairspray and air-freshener carry the label 'ozone friendly'. What does that tell you?

Questions on Section 7

1 a What does the term *electrolysis* mean?
 b Copy the diagram below, and label it using the words in this list, which are all connected with electrolysis:
anode, cathode, electrolyte, anion, cation.

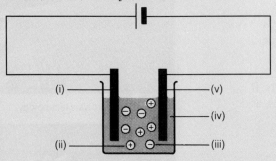

2 In which of these would the bulb light?

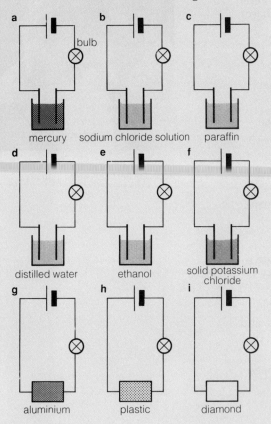

3 a Which of the substances in question 2 are:
  i conductors? ii non-conductors?
  iii electrolytes? iv non-electrolytes?
 b What is the difference between a conductor and an electrolyte?
 c For which substances above would you expect to see changes taking place at the electrodes?

4 The electrolysis of lead bromide can be investigated using the following apparatus.

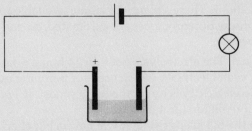

 a What must be done to the lead bromide before the bulb will light?
 b What would be *seen* at the positive electrode during the experiment?
 c Name the substance in **b**.
 d What is formed at the negative electrode?
 e Write an equation for the reaction at each electrode.

5 This question is about the electrolysis of *molten* lithium chloride. Lithium chloride is ionic, and contains lithium ions (Li^+) and chloride ions (Cl^-).
 a Which ion is the anion?
 b Which ion is the cation?
 c Copy the following diagram and use arrows to show which way:
  i the ions flow when the switch is closed;
  ii the electrons flow in the wires.

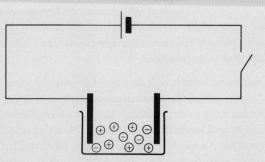

 d Write equations for the reaction at each electrode, and the overall reaction.

6 This question is about the electrolysis of an aqueous solution of lithium chloride.
 a Write down the names and symbols of all the ions present in the solution.
 b Lithium is a reactive metal, like sodium. What will be formed at the cathode?
 c What will be formed at the anode?
 d Write an equation for the reaction at each electrode.
 e Name two other electrolytes that will give the same electrolysis products as this one.

7 Write an equation for:
 a the overall decomposition
 b the reaction at each electrode
 when molten sodium chloride is electrolysed.

8 a List the anions and cations present in:
 i sodium chloride solution
 ii copper(II) chloride solution
 b Write down the reaction you would expect at:
 i the anode **ii** the cathode
 when each solution in **a** is electrolysed, using
 platinum electrodes.
 c Explain why the anode reactions in **b** are both the
 same.
 d Explain why copper is obtained at the cathode,
 but not sodium.

9 Six substances A to F were dissolved in water, and
 connected in turn into the circuit below. The
 symbol Ⓐ represents an ammeter, which is an
 instrument for measuring the current.

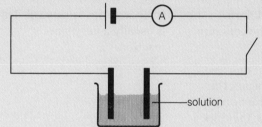

The results are shown in this table:

Substance	Current (amperes)	Cathode (+)	Anode (−)
A	0.8	copper	chlorine
B	1.0	hydrogen	chlorine
C	0.0	——	——
D	0.8	copper	oxygen
E	1.2	hydrogen	oxygen
F	0.7	silver	oxygen

a Which solution conducts best?
b Which solution is a non-electrolyte?
c Which solution could be:
 i silver nitrate?
 ii copper(II) sulphate?
 iii copper(II) chloride?
 iv sodium hydroxide?
 v sugar?
 vi potassium chloride?

10 Hydrogen chloride is a molecular substance.
 However, it dissolves in water to form *hydrochloric
 acid*, which exists as ions:
$$HCl\,(g) \longrightarrow H^+\,(aq) + Cl^-\,(aq)$$
 List the ions present in a solution of hydrochloric
 acid. What result would you expect, when the
 solution is electrolysed with platinum electrodes?

11 Aluminium is extracted by electrolysis of molten
 aluminium oxide. The aluminium ion is Al^{3+}.
 a Write an equation for the reaction at the cathode.
 b How many moles of electrons are required to
 obtain 1 mole of aluminium atoms?
 c What mass of aluminium will be obtained if a
 current of 25 000 A flows for 24 hours?
 (The RAM of Al = 27; the charge due to 1 mole of
 electrons is 96 500 coulombs.)

12 Using platinum electrodes the apparatus below was
 set up to electrolyse three different solutions.

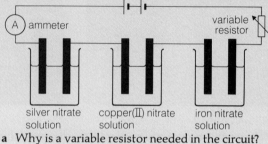

 a Why is a variable resistor needed in the circuit?
 b Write an electron transfer equation to show how
 silver is formed at the negative electrode.
 c In the experiment, 0.403 g of silver was formed
 when a current was passed for 30 minutes.
 The RAM of Ag = 108; the charge due to 1 mole
 of electrons is 96 500 coulombs.
 i Calculate the current used.
 ii Calculate the mass of copper formed in the
 middle cell. (The RAM of Cu = 64.)
 d In the third cell, 0.070 g of iron was obtained. The
 RAM of Fe = 56. Calculate whether the solution
 used was iron(II) nitrate (containing Fe^{2+} ions), or
 iron(III) nitrate (containing Fe^{3+} ions).
 e What would be obtained at the positive
 electrode in each beaker?

13 When sodium chloride solution is electrolysed, the
 gases hydrogen and chlorine are obtained.
 a i Write down the formulae for the hydrogen
 and chlorine ions, then say which gas is
 obtained at each electrode.
 ii Explain why hydrogen is released instead of
 sodium metal.
 b i Write the equation for the formation of
 chlorine gas at the electrode.
 ii How many moles of electrons are required to
 release 1 mole of chlorine gas (Cl_2)?
 iii How many coulombs is this? (The charge due
 to 1 mole of electrons is 96 500 coulombs.)
 c In a laboratory electrolysis, a current of 2 A was
 passed for 20 minutes.
 i Calculate the volume of chlorine released at
 room temperature and pressure (rtp).
 (The volume of 1 mole of any gas is 24 dm³ or
 24 000 cm³ at rtp.)
 ii Explain why the volume of hydrogen released
 is the same as the volume of chlorine.

8.1 Rates of reaction

Fast and slow

Some reactions are **fast** and some are **slow**. Look at these examples:

Silver chloride precipitating, when solutions of silver nitrate and sodium chloride are mixed. This is a very fast reaction.

Concrete setting. This reaction is quite slow. It will take a couple of days for the concrete to harden.

Rust forming on a heap of scrap iron in a scrapyard. This is usually a very slow reaction.

It is not always enough to know just that a reaction is fast or slow. For example, in a factory that makes products from chemicals, the chemical engineers need to know *exactly* how fast each reaction is going, and how long it takes to complete. In other words, they need to know the **rate** of each reaction.

What is rate?

Rate is a measure of how fast or slow something is. Here are some everyday examples:

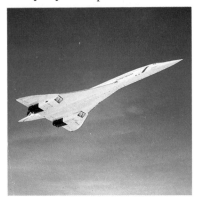

This plane has just flown 2000 kilometres in 1 hour. It flew at a **rate** of 2000 kilometres per hour.

This petrol pump can pump petrol at a **rate** of 50 litres per minute.

This machine can print newspapers at a **rate** of 10 copies per second.

From these examples you can see that:

Rate is a measure of the change that happens in a single unit of time.

Any suitable unit of time can be used – a second, a minute, an hour, even a day.

Rate of a chemical reaction

When zinc is added to dilute sulphuric acid, they react together. The zinc disappears slowly, and a gas bubbles off.

After a time, the bubbles of gas form less quickly. The reaction is slowing down.

Finally, no more bubbles appear. The reaction is over, because all the acid has been used up. Some zinc remains behind.

In this example, the gas that forms is hydrogen. The equation for the reaction is:

$$\text{zinc} + \text{sulphuric acid} \longrightarrow \text{zinc sulphate} + \text{hydrogen}$$
$$\text{Zn}\,(s) + \text{H}_2\text{SO}_4\,(aq) \longrightarrow \text{ZnSO}_4\,(aq) + \text{H}_2\,(g)$$

Both zinc and sulphuric acid get used up in the reaction. At the same time, zinc sulphate and hydrogen form.

You could measure the rate of the reaction, by measuring either:

● the amount of zinc used up per minute *or*
● the amount of sulphuric used up per minute *or*
● the amount of zinc sulphate produced per minute *or*
● the amount of hydrogen produced per minute.

For this reaction, it is easiest to measure the amount of hydrogen produced per minute – the hydrogen can be collected as it bubbles off, and its volume can then be measured. On the next page you will find out how this is done.

In general, to find the rate of a reaction, you should measure:
   the amount of a reactant used up per unit of time *or*
   the amount of a product produced per unit of time.

Questions

1 Here are some reactions that take place in the home. Put them in order of decreasing rate (the fastest one first).
   a Gloss paint drying.
   b Fruit going rotten.
   c Cooking gas burning.
   d A cake baking.
   e A metal bath rusting.
2 Which of these rates of travel is slowest?
   5 kilometres per second.
   20 kilometres per minute.
   60 kilometres per hour.

3 Suppose you had to measure the rate at which zinc is used up in the reaction above. Which of these units would be suitable?
   a litres per minute;
   b grams per minute;
   c centimetres per minute.
   Explain your choice.
4 Iron reacts with sulphuric acid like this:
   $$\text{Fe}\,(s) + \text{H}_2\text{SO}_4\,(aq) \longrightarrow \text{FeSO}_4\,(aq) + \text{H}_2\,(g)$$
   a Write a word equation for this reaction.
   b Write down four different ways in which the rate of the reaction could be measured.

8.2 Measuring the rate of a reaction

On the last page you saw that the rate of a reaction is found by measuring the amount of a **reactant** used up per unit of time or the amount of a **product** produced per unit of time.
Take, for example, the reaction between magnesium and excess dilute hydrochloric acid. Its equation is:

$$\text{magnesium} + \text{hydrochloric acid} \longrightarrow \text{magnesium chloride} + \text{hydrogen}$$
$$Mg\,(s) \quad + \quad 2HCl\,(aq) \quad \longrightarrow \quad MgCl_2\,(aq) \quad + \quad H_2\,(g)$$

In this reaction, hydrogen is the easiest substance to measure. This is because it is the only gas in the reaction. It bubbles off and can be collected in a **gas syringe**, where its volume is measured.

Some reactions are so fast that their rates would be very difficult to measure, like this detonation of an old mine.

The method   This apparatus is suitable:

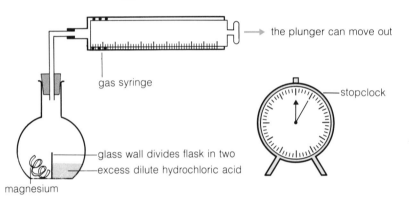

the plunger can move out

gas syringe

stopclock

glass wall divides flask in two
excess dilute hydrochloric acid

magnesium

The magnesium is cleaned with sandpaper and put into one part of the flask. Dilute hydrochloric acid is put into the other part. The flask is tipped up to let the two reactants mix, and the clock is started at the same time. Hydrogen begins to bubble off. It rises up the flask, and pushes its way into the gas syringe. The plunger is forced to move out:

At the start the plunger is fully in. No gas has yet been collected.

Now the plunger has moved out to the 20 cm³ mark. 20 cm³ of gas been been collected.

The volume of gas in the syringe is noted at intervals, for example at the end of each half-minute. How will you know when the reaction is complete?

The results   Here are some typical results:

Time/minutes	0	½	1	1½	2	2½	3	3½	4	4½	5	5½	6	6½
Volume of hydrogen/cm³	0	8	14	20	25	29	33	36	38	39	40	40	40	40

These results can be plotted on a graph, as shown on the next page.

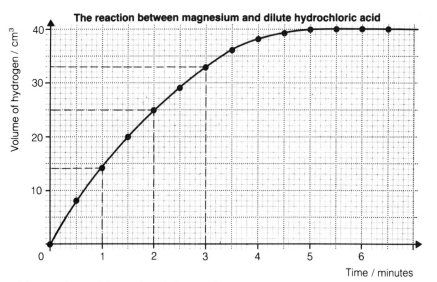

The reaction between magnesium and dilute hydrochloric acid

Notice these things about the results:

1. In the first minute, 14 cm³ of hydrogen is produced.
   So the rate for the first minute is 14 cm³ of hydrogen per minute.
   In the second minute, only 11 cm³ is produced. (25 − 14 = 11)
   So the rate for the second minute is 11 cm³ of hydrogen per minute.
   The rate for the third minute is 8 cm³ of hydrogen per minute.
   So you can see that the rate decreases as time goes on.
   The rate changes all through the reaction. It is greatest at the start, but gets less as the reaction proceeds.

2. The reaction is fastest in the first minute, and the curve is steepest then. It gets less steep as the reaction gets slower.
   The faster the reaction, the steeper the curve.

3. After 5 minutes, no more hydrogen is produced, so the volume no longer changes. The reaction is over, and the curve goes flat.
   When the reaction is over, the curve goes flat.

4. Altogether, 40 cm³ of hydrogen is produced in 5 minutes.

   The *average* rate for the reaction $= \dfrac{\text{total volume of hydrogen}}{\text{total time for the reaction}}$

   $= \dfrac{40\ \text{cm}^3}{5\ \text{minutes}}$

   $=$ **8 cm³ of hydrogen per minute.**

Note that this method can be used to measure the rate of *any* reaction in which one product is a gas — like the reaction shown on page 113.

curve flat, reaction over
curve less steep, reaction slower
curve steepest, reaction fastest

Questions

1. For this experiment, can you explain why:
   a. a divided flask is used?
   b. the magnesium ribbon is first cleaned?
   c. the clock is started the moment the reactants are mixed?
2. From the graph above, how can you tell when the reaction is over?

3. This question is about the graph above.
   a. How much hydrogen is produced in:
      i. 2.5 minutes?   ii. 4.5 minutes?
   b. How many minutes does it take to produce:
      i. 10 cm³   ii. 20 cm³   of hydrogen?
   c. What is the rate of the reaction during:
      i. the fourth minute?   ii. the fifth minute?

8.3 Changing the rate of a reaction (I)

The effect of concentration

A reaction can be made to go faster or slower by changing the
concentration of a reactant.

Suppose the experiment with magnesium and excess hydrochloric
acid is repeated twice (A and B below). Everything is kept the same
each time, *except* the concentration of the acid:

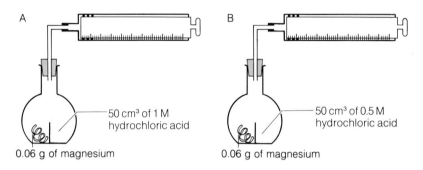

A   50 cm³ of 1 M
hydrochloric acid

0.06 g of magnesium

B   50 cm³ of 0.5 M
hydrochloric acid

0.06 g of magnesium

The acid in A is *twice as concentrated* as the acid in B.
Here are both sets of results shown on the same graph.

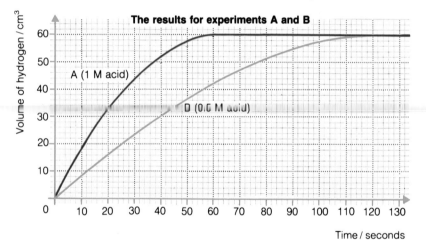

The results for experiments A and B

Volume of hydrogen / cm³

A (1 M acid)

B (0.5 M acid)

Time / seconds

Notice these things about the results:

1 Curve A is steeper than curve B. From this you can tell straight
   away that the reaction was faster in A than in B.
2 In A, the reaction lasts 60 seconds. In B it lasts for 120 seconds.
3 Both reactions produced 60 cm³ of hydrogen. Do you agree?
   In A it was produced in 60 seconds, so the average rate was 1 cm³
   of hydrogen per second. In B it was produced in 120 seconds, so
   the average rate was 0.5 cm³ of hydrogen per second.
   The average rate in A was twice the average rate in B.

These results show that:

**A reaction goes faster when the concentration of a reactant is
increased.**

For the reaction above, the rate doubles when the concentration of
acid is doubled.

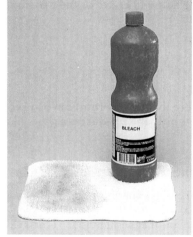

This stain could be removed with a
solution of bleach. The more
concentrated the solution, the faster the
stain will disappear.

The effect of temperature

A reaction can also be made to go faster or slower by changing the **temperature** of the reactants.

This time, a different reaction is used: when dilute hydrochloric acid is mixed with sodium thiosulphate solution, a fine yellow precipitate of sulphur forms. The rate can be followed like this:

1 A cross is marked on a piece of paper.
2 A beaker containing some sodium thiosulphate solution is put on top of the paper. The cross should be easy to see through the solution, from above.
3 Hydrochloric acid is added quickly, and a clock started at the same time. The cross grows fainter as the precipitate forms.
4 The clock is stopped the moment the cross can no longer be seen from above.

View from above the beaker:

The cross grows fainter with time

The experiment is repeated several times. The quantity of each reactant is kept exactly the same each time. Only the temperature of the reactants is changed. This table shows the results:

Temperature/°C	20	30	40	50	60
Time for cross to disappear/seconds	200	125	50	33	24

The higher the temperature, the faster the cross disappears

The cross disappears when enough sulphur forms to blot it out. Notice that this takes 200 seconds at 20 °C, but only 50 seconds at 40 °C. So the reaction is *four times faster* at 40 °C than at 20 °C.
A reaction goes faster when the temperature is raised. When the temperature increases by 10 °C, the rate approximately doubles.
This fact is used a great deal in everyday life. For example, food is kept in the fridge to slow down decomposition reactions and keep it fresh for longer. Can you think of any other examples?

The low temperature in the fridge slows down decomposition reactions.

Questions

1 Look at the graph on the opposite page.
   a After two minutes, how much hydrogen was produced in:
       i experiment A?   **ii** experiment B?
   b From the shape of the curves, how can you tell which reaction was faster?
2 Explain why experiments A and B both produce the same amount of hydrogen.

3 Copy and complete: A reaction goes . . . . when the concentration of a . . . . is increased. It also goes . . . . when the . . . . is raised.
4 Why does the cross disappear, in the experiment with sodium thiosulphate and hydrochloric acid?
5 What will happen to the rate of a reaction when the temperature is *lowered*? Use this to explain why milk is stored in a fridge.

8.4 Changing the rate of a reaction (II)

The effect of surface area

In many reactions, one of the reactants is a solid. The reaction between hydrochloric acid and calcium carbonate (marble chips) is one example. Carbon dioxide gas is produced:

$$CaCO_3 + 2HCl\,(aq) \longrightarrow CaCl_2\,(aq) \longrightarrow + H_2O\,(l) + CO_2\,(g)$$

The rate can be measured using the apparatus on the right.

The method Marble chips and acid are placed in the flask, which is then plugged with cotton wool. This prevents any liquid from splashing out during the reaction. Next the flask is weighed. Then it is tipped up, to let the reactants mix, and a clock is started at the same time. The mass is noted at regular intervals, until the reaction is complete.

Since carbon dioxide can escape through the cotton wool, the flask gets lighter as the reaction proceeds. So by weighing the flask, you can follow the rate of the reaction.

The experiment is repeated twice. Everything is kept exactly the same each time, except the **surface area** of the marble chips:

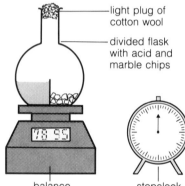

light plug of cotton wool

divided flask with acid and marble chips

balance   stopclock

For experiment 1, large chips of marble are used. The surface area is the total area of the surface of these chips.

For experiment 2, the same *mass* of marble is used. But this time it is in small chips, so its surface area is greater.

The results The results of the two experiments are plotted below:

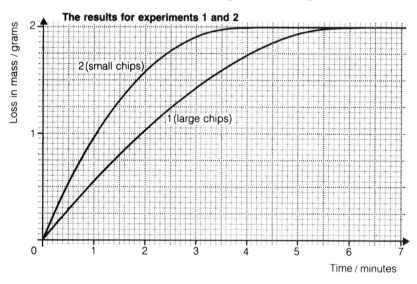

The results for experiments 1 and 2

Loss in mass / grams

2 (small chips)

1 (large chips)

Time / minutes

> **How to draw the graph:**
>
> First you have to find the *loss in mass* at different times:
>   loss in mass at a given time =
>   mass at start − mass at that time
> Then you plot the values for loss in mass against the times.

You should notice these things about the results:

1 Curve 2 is steeper than curve 1. This shows immediately that the reaction is faster for the small chips.
2 In both experiments, the final loss in mass is 2.0 grams. In other words, 2.0 grams of carbon dioxide is produced each time.
3 For the small chips, the reaction is complete in 4 minutes. For the large chips, it lasts for 6 minutes.

These results show that:

The rate of a reaction increases when the surface area of a solid reactant is increased.

The effect of a catalyst

Hydrogen peroxide is a clear, colourless liquid with the formula H_2O_2. It can decompose to water and oxygen:

hydrogen peroxide $\longrightarrow$ water + oxygen
$$2H_2O_2\,(aq) \longrightarrow 2H_2O\,(l) + O_2\,(g)$$

The rate of the reaction can be followed by collecting the oxygen:

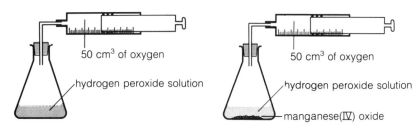
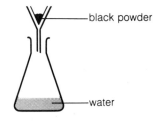

The reaction is in fact very slow. It could take 500 days to collect 50 cm³ of oxygen.

However, if 1 gram of black manganese(IV) oxide is added, the reaction goes much faster. 50 cm³ of oxygen is produced in a few minutes.

After the reaction, the black powder is removed by filtering. It is then dried, weighed and tested. It is still manganese(IV) oxide, and still weighs 1 gram.

The manganese(IV) oxide speeds up the reaction without being used up itself. It is called a **catalyst** for the reaction.

A catalyst is a substance that changes the rate of a chemical reaction but remains chemically unchanged itself.

Catalysts have been discovered for many reactions. They are usually **transition metals** or **compounds of transition metals.** For example, iron speeds up the reaction between nitrogen and hydrogen, to make ammonia (page 166). There are also many biological catalysts, called **enzymes**. For example, the pancreatic juice made by your pancreas contains enzymes that speed up digestion.

Questions

1 This question is about the graph on the opposite page. For each experiment find:
   a the loss in mass in the first minute;
   b the mass of carbon dioxide produced during the first minute;
   c the average rate of production of the gas.

2 What is a catalyst? Give two examples of reactions and their catalysts.

3 Look again at the decomposition of hydrogen peroxide, above. How would you show that:
   a the reaction goes even faster *if more than* 1 gram of catalyst is used?   b the catalyst is not used up?

8.5 Explaining rates

A closer look at a reaction

On page 114, you saw that magnesium and dilute hydrochloric acid react together:

magnesium + hydrochloric acid ⟶ magnesium chloride + hydrogen
$$Mg\,(s) \quad + \quad 2HCl\,(aq) \quad \longrightarrow \quad MgCl_2\,(aq) \quad + \quad H_2\,(g)$$

In order for the magnesium and acid particles to react together:
- **they must collide with each other.**
- **the collision must have enough energy.**

This is shown by the drawings below.

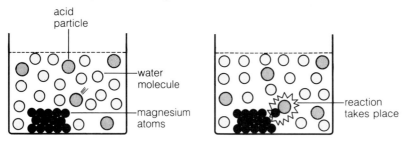

The particles in the liquid move around continually. Here an acid particle is about to collide with a magnesium atom.

If the collision has enough energy, reaction takes place. Magnesium chloride and hydrogen are formed.

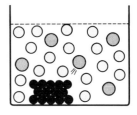

If the collision does not have enough energy, no reaction occurs. The acid particle bounces away again.

If there are lots of successful collisions in a given minute, then a lot of hydrogen is produced in that minute. In other words, the reaction goes quickly – its rate is high. If there are not many, its rate is low. **The rate of a reaction depends on how many successful collisions there are in a given unit of time.**

Changing the rate of a reaction

Why rate increases with concentration   If the concentration of the acid is increased, the reaction goes faster. It is easy to see why:

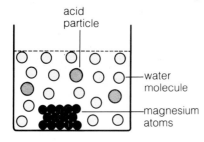
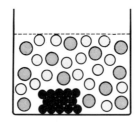

In dilute acid, there are not so many acid particles. This means there is not much chance of an acid particle hitting a magnesium atom.

Here the acid is more concentrated – there are more acid particles in it. There is now more chance of a successful collision occurring.

The more successful collisions there are, the faster the reaction.

This idea also explains why the reaction between magnesium and hydrochloric acid slows down as time goes on:

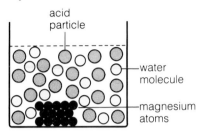

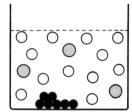

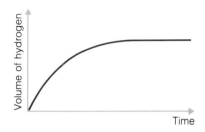

At the start, there are plenty of magnesium atoms and acid particles. But they get used up during successful collisions.

After a time, there are fewer magnesium atoms, and the acid is less concentrated. So the reaction slows down.

This means that the slope of the reaction curve decreases with time, as shown above.

Why rate increases with temperature At low temperatures, particles of reacting substances do not have much energy. However, when the substances are heated, the particles take in energy. This causes them to move faster and collide more often. The collisions have more energy, so more of them are successful. Therefore the rate of the reaction increases.

Why rate increases with surface area The reaction between the magnesium and acid is much faster when the metal is powdered:

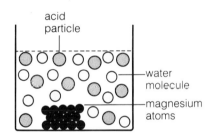

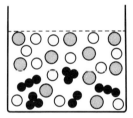

The acid particles can collide only with those magnesium atoms in the outer layer of the metal ribbon.

When the metal is powdered, many more atoms are exposed. So there is a greater chance of successful collisions.

Why a catalyst increases the rate Some reactions can be speeded up by adding a catalyst. *In the presence of a catalyst, a collision needs less energy in order to be successful.* The result is that more collisions become successful, so the reaction goes faster. Catalysts are very important in industry, because they speed up reactions even at low temperatures. This means that less fuel is needed, so money is saved.

Some catalysts used in industry.

Questions

1 Copy and complete: Two particles can only react together if they . . . . and if the . . . . has enough . . . .
2 What is:
  a a successful collision?
  b an unsuccessful collision?

3 In your own words, explain why the reaction between magnesium and acid goes faster when:
  a the temperature is raised;
  b the magnesium is powdered.
4 Explain why a catalyst can speed up a reaction, even at low temperatures.

8.6 Reversible reactions

You saw how you can speed up a reaction using heat, a catalyst and so on. But reactions aren't always so simple. Look at this one:

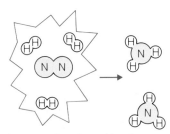 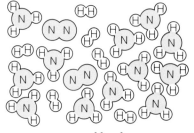 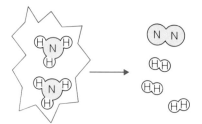

| Three molecules of hydrogen react with one of nitrogen to form two of ammonia. So you'd think that when . . . | . . . nitrogen and hydrogen are mixed in the right amounts, it would all turn into ammonia. But that never happens! | After a certain point, every time two ammonia molecules form, another two break down into nitrogen and hydrogen. |

The reaction between the nitrogen and hydrogen never gets completed because there are two competing reactions going on:

$$N_2(g) + 3H_2(g) \longrightarrow 2NH_3(g) \quad \text{and} \quad 2NH_3(g) \longrightarrow N_2(g) + 3H_2(g)$$
$$\text{the \textbf{forward} reaction} \qquad\qquad\qquad \text{the \textbf{back} reaction}$$

So the reaction is a **reversible reaction**. It is usually written like this:

$$N_2(g) + 3H_2(g) \rightleftharpoons 2NH_3(g)$$

The symbol $\rightleftharpoons$ indicates a reversible reaction.

Equilibrium

Once the ammonia in the mixture above reaches a certain level, the forward and back reactions balance each other. Every time two ammonia molecules form, another two break down. So the rate of the forward reaction is exactly the same as the rate of the back reaction. The reaction is in **equilibrium.**
A reversible reaction is in equilibrium when the forward and back reactions proceed at the same rate.

The problem with equlibrium . . .

Once the reaction above reaches equilibrium, the level of ammonia in the mixture stays steady. So if you run an ammonia factory, you have a problem. The more ammonia you make, the more you can sell. How can you improve the yield of ammonia, and so improve your profits?

This idea will help you:
When a reversible reaction is in equilibrium and you make a change, it will do what it can to oppose that change.

You can't make the reaction go to completion. A reversible reaction *always* ends up in a state of equilibrium. But by changing the conditions you can *shift the equilibrium to the right*, to give more ammonia in the mixture. So let's see what changes you can make.

Nitrogen, hydrogen, and ammonia in equilibrium. The mixture is in balance:

By changing the reaction conditions, you can shift equilibrium to the right, giving more ammonia:

1 Increasing the temperature

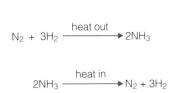

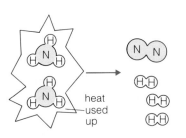

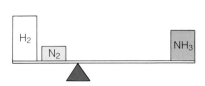

As you saw, heat speeds up *any* reaction. But here the forward reaction is exothermic – it gives out heat. The back reaction is endothermic – it takes it in.

If you heat the equilibrium mixture it will act to oppose the change. More ammonia will , break down in order to use up the heat you've added.

The result is that the reaction reaches equilibrium faster, which is good, but the level of ammonia *decreases*. So you're worse off than before.

2 Increasing the pressure

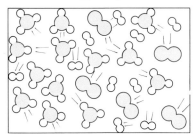

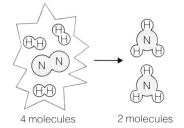

4 molecules   2 molecules

Pressure is caused by collisions between gas molecules and the walls of the container. So the fewer the molecules present, the lower the pressure.

If you apply more pressure, the equilibrium mixture will act to oppose the change. More ammonia will form in order to reduce the number of molecules.

The result is that the level of ammonia *increases*. Equilibrium shifts to the right. Well done. You're on the right track.

3 Removing the ammonia   The equilibrium mixture is a balance between the levels of nitrogen, hydrogen, and ammonia present. Suppose you cool the mixture. Ammonia condenses before nitrogen and hydrogen do, so you could run it off as a liquid. Then warm the remaining nitrogen and hydrogen again, to get them to react again. *Et voila*, more ammonia!

When you remove ammonia, nitrogen and hydrogen react to restore equilibrium.

4 Adding a catalyst   Iron acts as a catalyst for this reaction. But it speeds up the forward and back reactions *equally*. So the reaction reaches equilibrium faster, but the equilibrium position does not change. Still, it's worth using a catalyst because it saves you time, and therefore money.

So, for the best yield of ammonia, you should:
- use moderate heat and a catalyst, to reach equilibrium quickly.
- choose high pressure, and remove ammonia, to shift equilibrium right.

See how these conditions are applied in real life, on page 166.

Questions

1 The reaction between nitrogen and hydrogen is *reversible*. Explain what that means.

2 In manufacturing ammonia: **a** high pressure is used;   **b** ammonia is removed.   Explain why.

8.7 More about catalysts

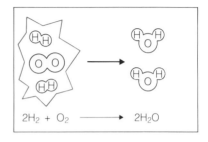

This gas jar contains a mixture of hydrogen and oxygen. Even if you leave it for hours, the two gases won't react together.

But hold a platinum wire in the mouth of the jar, and the gases explode immediately with a pop, producing water.

The reaction is exothermic. When 2 moles of hydrogen gas reacts with 1 mole of oxygen gas, 486 kJ of energy is given out.

So platinum is a **catalyst** for the reaction between hydrogen and oxygen. It speeds up the reaction, while remaining chemically unchanged itself. Look at these energy diagrams to see how it works:

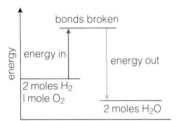

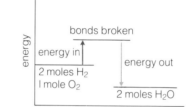

For the gases to react, bonds must first be broken. In the first jar above, there isn't enough energy for bonds to break. So the reaction rate is zero.

When platinum is present, less energy is needed to break bonds. This means that some molecules already have enough energy to break up . . .

. . . into atoms, and start the reaction. Once this happens, the energy given out breaks further bonds. The reaction goes so fast it's explosive.

So catalysts lower the energy needed for reaction to occur.

A closer look at the catalyst In the reaction above, platinum acts as a **surface catalyst**:

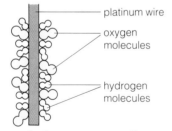

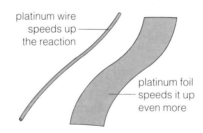

Instead of meeting on collision, gas molecules are adsorbed on to the metal surface. They get very close together, making it much easier for reaction to take place.

The larger its surface area, the better the platinum works, because more gas can be adsorbed. Many other catalysts work in the same way. For example . . .

. . . in catalytic converters in car exhausts, harmful gases are adsorbed on to a rhodium catalyst, where they react with oxygen to make less harmful products.

Enzymes

hydrogen
peroxide
solution

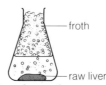
bubbles
of oxygen
manganese(IV)
oxide

froth
raw liver

cooked
liver

Hydrogen peroxide decomposes very slowly into water and oxygen.

Manganese(IV) oxide speeds up the reaction, making it thousands of times faster.

Raw liver also speeds it up. The liquid froths as the oxygen bubbles off.

If you boil the liver first, however, it will no longer speed up the reaction.

So something in the raw liver acts as a catalyst for this reaction. It is a protein called **catalase**. It is an example of an **enzyme**. **Enzymes are proteins which act as biological catalysts.**

We have hundreds of different enzymes in our bodies. For example the enzyme **amylase** in saliva breaks down starch into sugar in seconds. Without enzymes, reactions would be too slow at body temperature to keep us alive.

They are also important in industry. Enzymes in yeast speed up the fermentation of wine and beer, and the 'swelling' of dough for bread. Enzymes are used in making cheese and tenderizing meat. Enzymes that can break down proteins are used in biological washing powders, for removing stains such as blood and gravy.

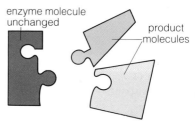
Without biological catalysts there would be no honey!

How enzymes work This shows an enzyme catalysing a decomposition reaction:

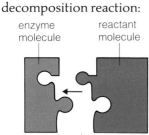
enzyme
molecule
reactant
molecule

reactant molecule
breaking down

enzyme molecule
unchanged
product
molecules

First, the molecules fit together like jigsaw pieces. For this, the reactant molecule has to be the right shape.

The 'complex' that forms makes it easier for the reactant molecule to break down. When decomposition is complete . . .

. . . the product molecules break away. Another molecule of reactant will take their place . . . as long as its shape is right!

Note Enzymes are different from other catalysts in these ways.

1 Because the shape of the molecules is so important, an enzyme will catalyse only a specific reaction. (But catalysts such as platinum can be used for many different reactions.)
2 If they are heated above about 45 °C, enzymes lose their shape and stop working. They become **denatured.**
3 You can also denature them by mixing them with acids and alkalis.

Questions

1 Which does a catalyst *not* change?
   a The speed of a reaction. **b** The products formed. **c** The amount of each product formed.
2 Explain how surface catalysts work.

3 Explain why: **a** boiled liver does not catalyse the decomposition of hydrogen peroxide;
   b you would not use a biological detergent on woollen sweater.

Questions on Section 8

1 The rate of the reaction between magnesium and dilute hydrochloric acid could be measured using this apparatus:

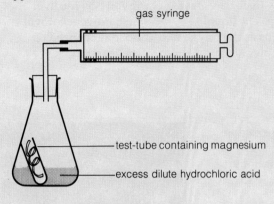

gas syringe

test-tube containing magnesium

excess dilute hydrochloric acid

a What is the purpose of:
   i the test-tube?
   ii the gas syringe?
b How would you get the reaction to start?

2 Some magnesium and an *excess* of dilute hydrochloric acid were reacted together. The volume of hydrogen produced was recorded every minute, as shown in the table:

Time/min	0	1	2	3	4	5	6	7
Volume of hydrogen/cm³	0	14	23	31	38	40	40	40

a What does an *excess* of acid mean?
b Plot a graph of the results, labelling the axes as on page 115.
c How much hydrogen was produced in:
   i the first minute?
   ii the second minute?
   iii the third minute?
   iv the fourth minute?
   v the fifth minute?
d What is the *rate of the reaction* (cm³ of hydrogen per minute) during each minute?
e What is the total volume of hydrogen produced in the reaction?
f How many minutes pass before the reaction finishes?
g What is the *average rate* of the reaction?
h A similar reaction had a rate of 15 cm³ of hydrogen in the first minute. Is this a slower or faster reaction than the one above?
i How could you make the above reaction go slower, while still using the same quantities of metal and acid?

3 For this question you will need the graph you drew for question 2.
The experiment with magnesium and an excess of dilute hydrochloric acid was repeated. This time a different concentration of hydrochloric acid was used. The results were:

Time/min	0	1	2	3	4	5	6
Volume of hydrogen/cm³	0	22	34	39	40	40	40

a Plot these results on the graph you drew for question 2.
b Which reaction was faster? How can you tell?
c In which experiment was the acid more concentrated? Give a reason for your answer.
d The same volume of hydrogen was produced in each experiment. What does that tell you about the mass of magnesium used?

4 Name three factors that affect the rate of a reaction, and describe the effect of changing each factor.

5 Suggest a reason for each of the following observations:
a Magnesium powder reacts faster than magnesium ribbon, with dilute sulphuric acid.
b Hydrogen peroxide decomposes much faster in the presence of the enzyme *catalase*.
c The reaction between manganese carbonate and dilute hydrochloric acid speeds up when some concentrated hydrochloric acid is added.
d Zinc powder burns much more vigorously in oxygen than zinc foil does.
e The reaction between sodium thiosulphate and hydrochloric acid takes a very long time if carried out in an ice bath.
f Zinc and dilute sulphuric acid react much more quickly when a few drops of copper(II) sulphate solution are added.
g Drenching with water prevents too much damage from spilt acid.
h A car's exhaust pipe will rust faster if the car is used a lot.
i In fireworks, powdered magnesium is used rather than magnesium ribbon.
j In this country, dead animals decay quite quickly. But in Siberia, bodies of mammoths that died 30 000 years ago have been found fully preserved in ice.
k The more sweet things you eat, the faster your teeth decay.
l Food cooks much faster in a pressure cooker than in an ordinary saucepan.

6 When sodium thiosulphate reacts with hydrochloric acid, a precipitate forms. In an investigation, the time taken for the solution to become opaque was recorded. (*Opaque* means that you cannot see through it.)

Four experiments (A to D) were carried out. Only the concentration of the sodium thiosulphate solution was changed each time.

The following results were obtained:

Experiment	A	B	C	D
Time taken/seconds	42	71	124	63

a Draw a diagram of suitable apparatus for this experiment.
b Name the precipitate that forms.
c What would be *observed* during the experiment?
d In which experiment was the reaction:
   i fastest? ii slowest?
e In which experiment was the sodium thiosulphate solution most concentrated? How can you tell?
f Suggest two other ways of speeding up this reaction.

7 Copper(II) oxide catalyses the decomposition of hydrogen peroxide. 0.5 g of the oxide was added to a flask containing 100 cm³ of hydrogen peroxide solution. A gas was released. It was collected and its volume noted every 10 seconds. This table shows the results:

Time/sec	0	10	20	30	40	50	60	70	80	90
Volume/cm³	0	18	30	40	48	53	57	58	58	58

a What is a catalyst?
b Draw a diagram of suitable apparatus for this experiment.
c Name the gas that is formed.
d Write a balanced equation for the decomposition of hydrogen peroxide.
e Plot a graph of the volume of gas (vertical axis) against time (horizontal axis).
f Describe how the rate changes during the reaction.
g What happens to the concentration of hydrogen peroxide as the reaction proceeds?
h What chemicals are present in the flask after 90 seconds?
i What mass of copper(II) oxide would be left in the flask at the end of the reaction?
j Sketch on your graph the curve that might be obtained for 1.0 g of copper(II) oxide.
k Name one other chemical that catalyses this decomposition.

8 When chlorine gas is passed over iodine, the following reactions take place.

Reaction 1: $I_2(s) + Cl_2(g) \longrightarrow 2\,ICl(l)$
                                                    brown

Reaction 2: $ICl(l) + Cl_2(g) \rightleftharpoons ICl_3(s)$
                  brown   yellow

a What is the first change you would see as the chlorine is passed over the iodine?
b As more chlorine is passed, what further change would you see?
c Which of the reactions can be *reversed*?
d What change would you see as the chlorine gas supply is turned off? Explain your answer.

9 Hydrogen and bromine react together to form hydrogen bromide:
   $H_2(g) + Br_2(g) \rightleftharpoons 2HBr(g)$
a Which of the following will favour the formation of more hydrogen bromide?
   i   Adding more hydrogen.
   ii   Removing bromine.
   iii Removing the product as it is formed.
b Explain why increasing the pressure has no effect on the amount of product formed.

10 Potassium chlorate decomposes when heated, like this:
   $2KClO_3(s) \longrightarrow 2KCl(s) + 3O_2(g)$
a Write a word equation for the reaction.
b Manganese(IV) oxide acts as a catalyst for this reaction. What would you expect to happen if two test-tubes, one of potassium chlorate and the other a mixture of potassium chlorate and manganese(IV) oxide, were heated?
c Potassium chloride is soluble in water, and manganese(IV) oxide is insoluble. How could you show that the manganese(IV) oxide is not used up during the reaction?
d How many moles of oxygen gas are obtained from two moles of potassium chlorate?
e Will there be *more* or *less* oxygen produced when the catalyst is used? Explain your answer.

11 The enzyme *polyphenoxidase* is involved in the oxidation reaction that causes sliced apple to turn brown in air. Explain the following observations.
a When the apple is first cut open, the flesh is not brown.
b Browning is much slower when the sliced apple is placed in the fridge.
c No browning takes place if the apple is placed in boiling water for 30 seconds straight after slicing.
d Apple that has been pulped in a food mixer turns brown much faster than sliced apple does.
e The browning reaction does not take place if the sliced apple is dipped in lemon juice straight away.

9.1 Acids and alkalis

Acids

One important group of chemicals is called **acids**:

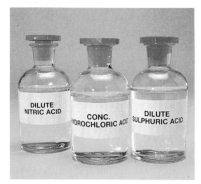

Here are some acids. You have probably seen them in the lab. They are all liquids. In fact they are **solutions** of pure compounds in water.

They must be handled carefully, especially the concentrated acids, for they are **corrosive**. They can eat away metals, skin, and cloth.

But some acids are not so corrosive, even when they are concentrated. They are called **weak** acids. Ethanoic acid is one example. It's found in vinegar.

You can tell if something is an acid, by its effect on **litmus**.
Litmus is a purple dye. It can be used as a solution, or on paper:

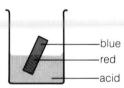

Litmus solution is purple. Litmus paper for testing acids is blue.

Acids will turn litmus solution red.

They will also turn blue litmus paper red.

Some common acids The main ones are:

hydrochloric acid	HCl (aq)
sulphuric acid	H_2SO_4 (aq)
nitric acid	HNO_3 (aq)
ethanoic acid	CH_3COOH (aq)

But there are plenty of others. For example, lemon juice contains **citric acid**, ant and nettle stings contain **methanoic acid** and tea contains **tannic acid**.

Alkalis

There is another group of chemicals that also affect litmus, but in a different way to acids. They are the **alkalis**.
Alkalis turn litmus solution blue, and red litmus paper blue.
Like acids, they must be handled carefully, because they can burn skin.

Some common alkalis Most pure alkalis are solids. But they are usually used in laboratory as aqueous solutions. The main ones are:

sodium hydroxide	$NaOH$ *(aq)*
potassium hydroxide	KOH *(aq)*
calcium hydroxide	$Ca(OH)_2$ *(aq)*
ammonia	NH_3 *(aq)*

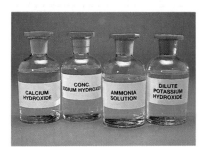

CALCIUM HYDROXIDE
CONC. SODIUM HYDROXIDE
AMMONIA SOLUTION
DILUTE POTASSIUM HYDROXIDE
Common laboratory alkalis.

Neutral substances

Many substances do not affect the colour of litmus solution, so they are not acids or alkalis. They are **neutral**. Sodium chloride and sugar solutions are both neutral.

The pH scale

You saw on page 128 that some acids are weaker than others. It is the same with alkalis.
The strength of an acid or an alkali is shown using a scale of numbers called the **pH scale**. The numbers go from 0 to 14:

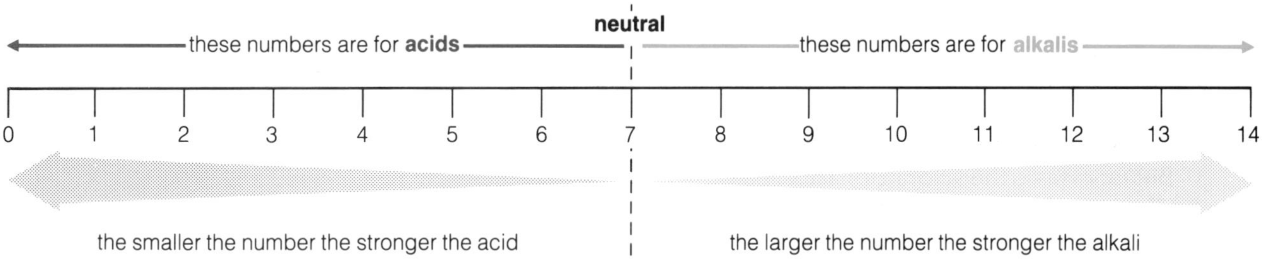

neutral
these numbers are for acids
these numbers are for alkalis
0 1 2 3 4 5 6 7 8 9 10 11 12 13 14
the smaller the number the stronger the acid
the larger the number the stronger the alkali

On this scale:
An acidic solution has a pH number less than 7.
An alkaline solution has a pH number greater than 7.
A neutral solution has a pH number of exactly 7.

You can find the pH of any solution by using **universal indicator**. Universal indicator is a mixture of dyes. Like litmus, it can be used as a solution, or as universal indicator paper. It goes a different colour at different pH values, as shown in this diagram:

red
orange
yellow
yellowish-green
green
greenish-blue
blue
violet
1 2 3 4 5 6 7 8 9 10 11 12 13

Questions

1. What does *corrosive* mean?
2. How would you test a substance, to see if it is an acid?
3. Write down the formula for:
   sulphuric acid — nitric acid
   calcium hydroxide — ammonia solution
4. What effect do alkalis have on litmus solution?
5. Say whether a solution is acidic, alkaline or neutral, if its pH number is:
   9 4 7 1 10 3
6. What colour would universal indicator show, in an aqueous solution of sugar? Why?

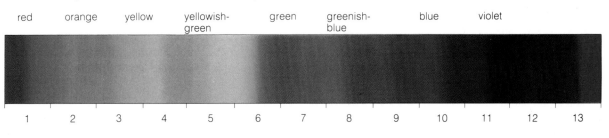

9.2 A closer look at acids

The liquids above are all **acidic**, as their pH numbers show. Which is the most acidic? How can you tell?

The properties of acids

1 Acids have a sour taste. Think of the taste of vinegar.
  But *never* taste the laboratory acids, because they could burn you.
2 They turn litmus red.
3 They have pH numbers less than 7.
4 They usually react with **metals**, forming hydrogen and a **salt**:

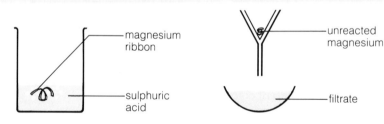

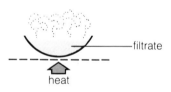

When magnesium is dropped into dilute sulphuric acid, hydrogen quickly bubbles off.

The bubbles stop when the reaction is over. The unreacted magnesium is then removed by filtering.

The filtrate is heated, to evaporate the water. A white solid is left behind. This solid is **magnesium sulphate**.

The equation for the reaction is:

magnesium + sulphuric acid $\longrightarrow$ magnesium sulphate + hydrogen
$Mg\,(s)$ + $H_2SO_4\,(aq)$ $\longrightarrow$ $MgSO_4\,(aq)$ + $H_2\,(g)$

The magnesium has driven the hydrogen out of the acid, and taken its place. Magnesium sulphate is called a **salt**.
When a metal takes the place of hydrogen in an acid, the compound that forms is called a salt.

The salts of sulphuric acid are always called **sulphates**.
The salts of hydrochloric acid are called **chlorides**.
The salts of nitric acid are called **nitrates**.

5 Acids react with **carbonates**, forming a salt, water and carbon dioxide. Hydrochloric acid reacts with calcium carbonate like this:

calcium carbonate + hydrochloric acid ⟶ calcium chloride + water + carbon dioxide

$$CaCO_3 \, (s) + 2HCl \, (aq) \longrightarrow CaCl_2 \, (aq) + H_2O \, (l) + CO_2 \, (g)$$

6 They react with **alkalis**, forming a salt and water. For example:

sodium hydroxide + nitric acid ⟶ sodium nitrate + water

$$NaOH \, (aq) + HNO_3 \, (aq) \longrightarrow NaNO_3 \, (aq) + H_2O \, (l)$$

7 They also react with metal oxides, forming a salt and water:

zinc oxide + hydrochloric acid ⟶ zinc chloride + water

$$ZnO \, (s) + 2HCl \, (aq) \longrightarrow ZnCl_2 \, (aq) + H_2O \, (l)$$

When an acid reacts with a carbonate, carbon dioxide fizzes off.

What causes acidity?

Acids have a lot in common, as you have just seen.
There must be something *in* them all, that makes them act alike. That 'something' is **hydrogen ions**.
Acids contain hydrogen ions.
Acids are solutions of pure compounds in water. The pure compounds are molecular. But in water, the molecules break up to form ions. They *always* give hydrogen ions. For example, in hydrochloric acid:

$$HCl \, (aq) \longrightarrow H^+ \, (aq) + Cl^- \, (aq)$$

The more H^+ ions there are in a solution, the more acidic it is. In other words the more H^+ ions there are, the lower the pH number.

Strong and weak acids

Dilute hydrochloric acid reacts quickly with magnesium ribbon. The reaction could be over in minutes. But when ethanoic acid of the same concentration is used instead, the reaction is much slower. It could take all day.

The hydrochloric acid reacts faster *because it contains more hydrogen ions.*
In hydrochloric acid, *nearly all* the acid molecules break up to form ions. It is called a **strong** acid. But in ethanoic acid, only some of the acid molecules form ions, so it is called a **weak** acid.
In a strong acid, nearly all the acid molecules form ions. In a weak acid, only some of the acid molecules form ions.
So a strong acid *always* has a lower pH number than a weak acid of the same concentration.
Some strong and weak acids are shown in the box on the right.

Dilute ethanoic acid reacts more slowly with magnesium than hydrochloric acid does, because it is a weak acid.

Strong acids
Hydrochloric acid
Sulphuric acid
Nitric acid

Weak acids
Ethanoic acid
Citric acid
Carbonic acid

Questions

1 Name the acid found in:
lemon juice   wine   tea   vinegar
2 Write a word equation for the reaction of dilute sulphuric acid with: **a** zinc
b magnesium oxide   **c** sodium carbonate

3 Bath salts contain sodium carbonate. Explain why they fizz when you put vinegar on them.
4 What causes the acidity in acids?
5 Why is ethanoic acid called a weak acid?
6 Name two other weak acids, and two strong ones.

9.3 A closer look at alkalis

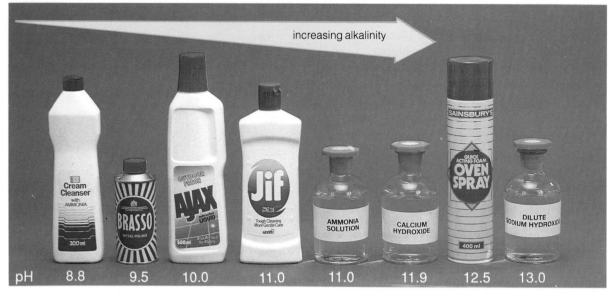

increasing alkalinity

| pH | 8.8 | 9.5 | 10.0 | 11.0 | 11.0 | 11.9 | 12.5 | 13.0 |

All the substances above are alkaline – you can tell by their pH numbers. Notice that several of them can be found in the kitchen! Many kitchen cleaners are alkaline because they contain ammonia or sodium hydroxide, which attack grease.

What makes things alkaline?

You saw on page 131 that all acid solutions contain hydrogen ions, H^+. The alkalis also have something in common:
All alkaline solutions contain hydroxide ions, OH^-.
In sodium hydroxide solution, the ions are produced like this:

$$NaOH\ (aq) \longrightarrow Na^+\ (aq) + OH^-\ (aq)$$

In ammonia solution, ammonia molecules react with water molecules to form ions:

$$NH_3\ (aq) + H_2O\ (l) \longrightarrow NH_4^+\ (aq) + OH^-\ (aq)$$

The more OH^- ions there are in a solution, the more alkaline it will be. In other words, the more OH^- ions there are, the higher the pH number.

Strong and weak alkalis

Like acids, alkalis can also be strong or weak.
Sodium hydroxide is a **strong** alkali, because it exists almost completely as ions, in solution. Potassium hydroxide is also a strong alkali. But ammonia solution is a **weak** alkali because only some ammonia molecules form ions in solution.

Properties of alkalis

1 Alkalis feel soapy to the touch. But it is dangerous to touch the laboratory alkalis (and some kitchen cleaners) because they can burn flesh.
2 Their solutions turn litmus blue.
3 Their solutions have pH numbers greater than 7.

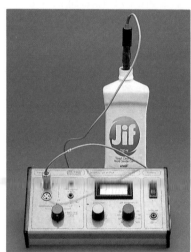

The pH number of a kitchen cleaner being measured with a pH meter.

Strong alkalis
Sodium hydroxide
Potassium hydroxide
Calcium hydroxide

Weak alkali
Ammonia

4 All the alkalis except ammonia will react with **ammonium compounds**. They drive ammonia out of the compounds, as a gas. For example:

$$\text{calcium hydroxide} + \text{ammonium chloride} \longrightarrow \text{calcium chloride} + \text{steam} + \text{ammonia}$$
$$\text{Ca(OH)}_2\,(s) + 2\text{NH}_4\text{Cl}\,(s) \longrightarrow \text{CaCl}_2\,(s) + 2\text{H}_2\text{O}\,(g) + 2\text{NH}_3\,(g)$$

This reaction is used for making ammonia in the laboratory.

5 All alkalis react with **acids**, as you saw on page 131, producing a salt and water.

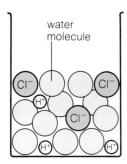

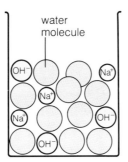

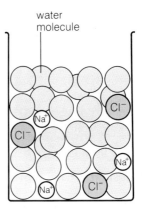

This is a solution of hydrochloric acid. It contains H$^+$ ions and Cl$^-$ ions. It will turn litmus red.

This is a solution of sodium hydroxide. It contains Na$^+$ ions and OH$^-$ ions. It will turn litmus blue.

When the two solutions are mixed, the H$^+$ and OH$^-$ ions join to form **water molecules**. The result is a neutral solution of sodium chloride, containing Na$^+$ and Cl$^-$ ions. It has no effect on litmus.

The equation tor this reaction is:

$$\text{HCl}\,(aq) + \text{NaOH}\,(aq) \longrightarrow \text{NaCl}\,(aq) + \text{H}_2\text{O}\,(l)$$

You could also write it showing only the ions that combine:

$$\text{H}^+\,(aq) + \text{OH}^-\,(aq) \longrightarrow \text{H}_2\text{O}\,(l)$$

The reaction is called a **neutralization**. The alkali has **neutralized** the acid by removing its H$^+$ ions, and turning them into water. **During a neutralization reaction, the H$^+$ ions of the acid are turned into water.**
A neutralization reaction is exothermic — it gives out heat. So the temperature of the solution rises a little.
You could obtain solid sodium chloride by heating the last solution above. The water evaporates leaving the solid behind.

Questions

1 Look at the substances shown at the top of page 132. Which ion do they all have in common?

2 What colour would litmus solution turn, if you mixed it with Ajax? Why?

3 A 0.1 M solution of potassium hydroxide has a higher pH number than a 0.1 M solution of ammonia. Why is this? Which one will be a better conductor of electricity?

4 Write a word equation for the reaction between sodium hydroxide and ammonium chloride.

5 What happens to:
a the H$^+$ ions **b** the temperature
of the solution when an acid is neutralized?

6 Write a balanced equation for the reaction between sodium hydroxide and:
a sulphuric acid **b** nitric acid

9.4 Bases and neutralization

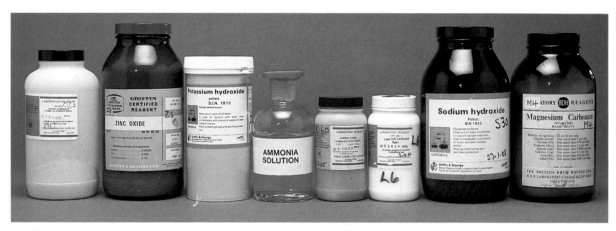

All these substances have something in common, as you will find out below.

Bases

An alkali can **neutralize** an acid, and destroy its acidity. It does this by removing the H^+ ions and converting them to water. But alkalis are not the only compounds that can neutralize acids:

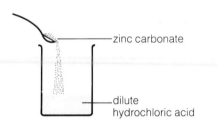

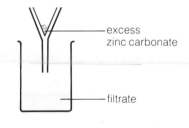

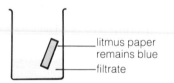

1 When zinc carbonate is added to dilute hydrochloric acid, it dissolves, and carbon dioxide bubbles off.

2 More is added until no more will dissolve, even with heating. The excess zinc carbonate is removed by filtering.

3 Then the filtrate is tested with blue litmus. There is no colour change. The acidity of the acid has been destroyed.

The equation for the reaction is:

$$ZnCO_3 \, (s) + 2HCl \, (aq) \longrightarrow ZnCl_2 \, (aq) + H_2O \, (l) + CO_2 \, (g)$$

So the H^+ ions of the acid have been turned into water. The acid has been neutralized by the zinc carbonate.

In fact, acid can be neutralized by any of these compounds:

> **metal oxides**
> **metal hydroxides**
> **metal carbonates**
> **metal hydrogen carbonates**
> **ammonia solution**

These compounds are all called **bases.**
Any compound that can neutralize an acid is called a base.

134

Alkalis are bases

The photograph at the top of the opposite page shows some bases. Notice that it includes alkalis – they can neutralize acids, so they are bases. They are bases that are soluble in water.
Alkalis are soluble bases.

Neutralization always produces a salt

Neutralization always produces a salt, as these general equations show:

acid + metal oxide $\longrightarrow$ metal salt + water

acid + metal hydroxide $\longrightarrow$ metal salt + water

acid + metal carbonate $\longrightarrow$ metal salt + water + carbon dioxide

acid + metal hydrogen carbonate $\longrightarrow$ metal salt + water + carbon dioxide

acid + ammonia solution $\longrightarrow$ ammonium salt + water

Some neutralizations in everyday life

Insect stings When a bee stings, it injects an acidic liquid into the skin. The sting can be neutralized by rubbing on **calamine lotion**, which contains zinc carbonate, or **baking soda**, which is sodium hydrogen carbonate.
Wasp stings are alkaline, and can be neutralized with vinegar. Why? Ant stings and nettle stings contain methanoic acid. How would you treat them?

Indigestion It may surprise you to know that you carry hydrochloric acid around in your stomach. It is a very dilute solution, and you need it for digesting food. But too much of it leads to **indigestion**, which can be very painful. To cure indigestion, you must neutralize the excess acid with a drink of sodium hydrogen carbonate solution (baking soda), or an indigestion tablet.

Soil treatment Most plants grow best when the pH of the soil is close to 7. If the soil is too acidic, or too alkaline, the plants grow badly or not at all.
Chemicals can be added to soil to adjust its pH. Most often, soil is too acid, so it is treated with **quicklime** (calcium oxide), **slaked lime** (calcium hydroxide), or **chalk** (calcium carbonate). These are all bases, and are quite cheap.

Factory waste Liquid waste from factories often contains acid. If it reaches a river, the acid will kill fish and other river life. This can be prevented by adding slaked lime to the waste, to neutralize it.

Bee stings are acidic.

Slaked lime being spread on fields where the soil is too acidic.

Questions

1 Look at the reaction shown on page 134. What salt does the filtrate contain? How would you obtain the dry salt from the filtrate?
2 What is a base? Name six bases.
3 What special property do alkalis have?
4 Which bases react with acids to give carbon dioxide?

5 Write a word equation for the reaction between:
   a dilute hydrochloric acid and copper(II) oxide;
   b dilute sulphuric acid and potassium carbonate.
6 Write a word equation for the reaction that takes place in your stomach, when you take baking soda to cure indigestion.

9.5 Making salts (I)

Making salts from acids

Acid + metal   Zinc sulphate can be made by reacting dilute sulphuric acid with zinc:

$$Zn\ (s) + H_2SO_4\ (aq) \longrightarrow ZnSO_4\ (aq) + H_2\ (g)$$

These are the steps:

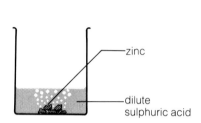 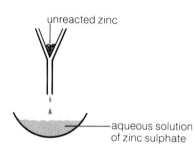

1 Some dilute sulphuric acid is put in a beaker, and zinc is added. The zinc begins to dissolve, and hydrogen bubbles off. The bubbles stop when all the acid has been used up.

2 Some zinc is still left. It is removed by filtering, which leaves an aqueous solution of zinc sulphate.

3 The solution is heated to evaporate some of the water. Then it is left to cool. Crystals of zinc sulphate start to form.

The method above is not suitable for *all* metals, or *all* acids.
It is fine for magnesium, aluminium, zinc and iron.
But the reactions of sodium, potassium and calcium with acid are dangerously violent. The reaction of lead is too slow, and copper, silver and gold do not react at all (page 146).

Acid + insoluble base   Copper(II) oxide is an insoluble base. Although copper will not react with dilute sulphuric acid, copper(II) oxide will. The salt that forms is copper(II) sulphate:

$$CuO\ (s) + H_2SO_4\ (aq) \longrightarrow CuSO_4\ (aq) + H_2O\ (l)$$

The method is quite like the one above.

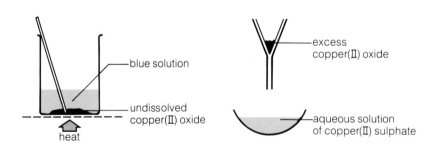 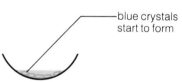

1 Some copper(II) oxide is added to dilute sulphuric acid. On warming, it dissolves and the solution turns blue. More is added until no more dissolves.

2 That means all the acid has been used up. The excess solid is removed by filtering. This leaves a blue solution of copper(II) sulphate in water.

3 The solution is heated to evaporate some of the water. Then it is left to cool. Blue crystals of copper(II) sulphate start to form.

Acid + alkali (soluble base) The reaction of sodium with acids is very dangerous. So sodium salts are usually made by starting with sodium hydroxide. This reaction can be used to make sodium chloride:

$$\text{NaOH}\,(aq) + \text{HCl}\,(aq) \longrightarrow \text{NaCl}\,(aq) + \text{H}_2\text{O}\,(l)$$

Both reactants are soluble, and no gas is given off during the reaction. So it is difficult to know when the reaction is over. You have to use an **indicator**. Universal indicator or litmus could be used, but even better is **phenolphthalein**. This is pink in alkaline solution, but colourless in neutral and acid solutions:

1 25 cm³ of sodium hydroxide solution is measured into a flask, using a pipette. Then two drops of phenolphthalein are added. The indicator turns pink.

2 The acid is added from a burette, a little at a time. The flask is swirled in a controlled way, to allow the acid and alkali to mix.

3 When all the alkali has been used up, the indicator suddenly turns colourless, showing that the solution is neutral. There is no need to add more acid.

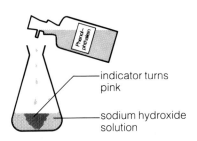

4 You can tell how much acid was added, using the scale on the burette. So now you know how much acid is needed to neutralize 25 cm³ of alkali.

5 The reaction is carried out again, but this time there is no need for an indicator. 25 cm³ of alkali is put in the flask, and the correct amount of acid added.

6 The solution from the flask is heated, to let the water evaporate. You will find dry crystals of sodium chloride are left behind.

In step 5 the reaction has to be carried out again, *without* indicator, because indicator would make the salt impure.
A similar method can be used for making potassium salts from potassium hydroxide, and ammonium salts from ammonia solution.

Questions

1 Name the acid and metal you would use for making:
   a zinc chloride   **b** magnesium sulphate
2 Why would you *not* make potassium chloride from potassium and hydrochloric acid?

3 How would you obtain lead(II) nitrate, starting with the insoluble compound lead(II) carbonate?
4 Write instructions for making potassium chloride, starting with solid potassium hydroxide.

9.6 Making salts (II)

Making insoluble salts

The salts made so far have all been soluble. They were obtained as crystals by evaporating solutions. But not all salts are soluble:

Soluble		Insoluble
All sodium, potassium, and ammonium salts		
All nitrates		
Chlorides...	*except*	silver and lead chloride
Sulphate...	*except*	calcium, barium and lead sulphate
Sodium, potassium, and ammonium carbonates...		but all other carbonates are insoluble

Insoluble salts can be made by **precipitation**.
For example, insoluble barium sulphate is precipitated when solutions of barium chloride and magnesium sulphate are mixed:

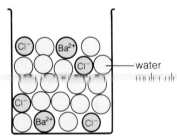

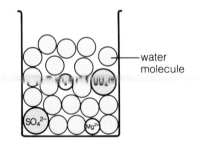

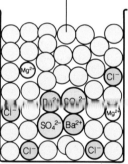

This is a solution of barium chloride, $BaCl_2$. It contains barium ions and chloride ions.

This is a solution of magnesium sulphate, $MgSO_4$. It contains magnesium ions and sulphate ions.

When the two solutions are mixed, the barium ions and sulphate ions bond together, because they are strongly attracted to each other. Solid barium sulphate precipitates.

The equation for the reaction is:

$$BaCl_2\,(aq) + MgSO_4\,(aq) \longrightarrow BaSO_4\,(s) + MgCl_2\,(aq)$$

or you could write it in a shorter way as:

$$Ba^{2+}\,(aq) + SO_4^{2-}\,(aq) \longrightarrow BaSO_4\,(s)$$

These are the steps for obtaining the barium sulphate:

1 Solutions of barium chloride and magnesium sulphate are mixed. A white precipitate of barium sulphate forms at once.
2 The mixture is filtered. The barium sulphate gets trapped in the filter paper.
3 It is rinsed with distilled water.
4 Then it is put in a warm oven to dry.

The precipitation of barium sulphate.

Barium sulphate could also be made from barium nitrate and sodium sulphate, for example, since these are both soluble. As long as barium ions and sulphate ions are present, barium sulphate will be precipitated.

To precipitate an insoluble salt, you must mix a solution that contains its positive ions with one that contains its negative ions.

Making salts in industry

Many salts occur naturally, and can be dug up out of the earth. Sodium chloride and magnesium sulphate are examples.
But others have to be made in factories, using methods like the ones on these pages. Here are some examples.

Fertilizers Huge amounts of nitrates, sulphates and phosphates are needed every year as fertilizers (page 168). They are made by neutralizing nitric acid, sulphuric acid and phosphoric acid in giant tanks.
One important fertilizer is **ammonium nitrate**. It is made by neutralizing nitric acid with ammonia solution:

$$HNO_3 \ (aq) + NH_3 \ (aq) \longrightarrow NH_4NO_3 \ (aq)$$

The water is driven off in an evaporator, leaving molten ammonium nitrate. This is sprayed down a tall tower. By the time it reaches the bottom it has cooled into small pellets that are easy to spread on soil.

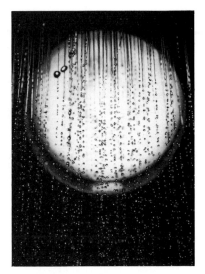

Molten ammonium nitrate being sprayed down a tower. It cools into pellets.

Soaps They might not look like it, but soaps are also salts! They are the sodium salts of the **fatty acids** found in animal fat and vegetable oil. Sodium hydroxide solution is added to the oil or melted fat, and the mixture is boiled. The acids are neutralized and soap is formed. It is purified, and perfume, colouring and other things added. Then it is pressed into cakes.

Pigments Pigments are insoluble coloured compounds used to colour paints and plastics. Many of them are insoluble salts made by precipitation. For example, **lead chromate** is reddish yellow and is made by mixing solutions of lead nitrate and sodium chromate.
Barium sulphate is white and is made by mixing solutions of barium chloride and magnesium sulphate, like in the photo on the opposite page, but on a much larger scale!

Soap moulds. Fat or oil, and sodium hydroxide are the main ingredients used in soap-making.

Questions

1 Choose two starting compounds you could use to precipitate: **a** calcium sulphate
   b zinc carbonate   **c** lead chloride
2 Write a balanced equation for each reaction in question 1.
3 Potassium sulphate is a fertilizer. How would you make it? Write an equation for the reaction.

4 Copy and complete:
   Salts are made in industry using the same methods you use in the lab. For example, fertilizers are made by . . . . an acid with a . . . . Soap is also made by . . . ., using . . . . and a . . . . from animal fat or vegetable oil. Pigments can be made by . . . . because they are insoluble.

Questions on Section 9

1 Match each solution from list A with the correct formula from list B.

List A (solutions)	List B (formulae)
Sodium hydroxide	H_2SO_4 (aq)
Hydrochloric acid	HNO_3 (aq)
Ammonia	HCl (aq)
Calcium hydroxide	CH_3COOH (aq)
Sulphuric acid	NH_3 (aq)
Nitric acid	$Ca(OH)_2$ (aq)
Ethanoic acid	$NaOH$ (aq)

2 Five solutions A to E were tested with universal indicator solution, to find their pH. The results are shown below.

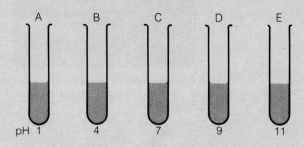

a What colour would each solution be?
b Which solution is:
   i neutral?
   ii strongly acidic?
   iii weakly acidic?
   iv strongly alkaline?
c The five solutions were known to be sodium chloride, sulphuric acid, ammonia solution, sodium hydroxide, and ethanoic acid. Now identify each of the solutions A to E.

3 Rewrite the following, choosing the correct word from each pair in brackets.
Acids are compounds which dissolve in water giving (hydrogen/hydroxide) ions. Sulphuric acid is one example. It is a (strong/weak) acid, which can be neutralized by (acids/alkalis) to form salts called (nitrates/sulphates).
Many (metals/nonmetals) react with acids to give a gas called (hydrogen/carbon dioxide). Acids also react with (chlorides/carbonates) to give (chlorine/carbon dioxide).
Solutions of acids are (good/poor) conductors of electricity. They also affect indicators. For example, phenolphthalein turns (pink/colourless) in acids, while litmus turns (red/blue).
The strength of an acid is shown by its (concentration/pH) number. The (higher/lower) the number, the stronger the acid.

4 State whether the following properties belong to acids, alkalis, or both.
a Sour taste.
b pH values greater than 7.
c Change the colour of litmus.
d Soapy to touch.
e Soluble in water.
f May be strong or weak.
g Neutralize bases.
h Form ions in water.
i Dangerous to handle.
j Form salts with certain other chemicals.
k Usually react with metals.

5 This is a brief description of a neutralization reaction.
'25 cm³ of potassium hydroxide solution was placed in a flask and a few drops of phenolphthalein were added. Dilute hydrochloric acid was added until the indicator changed colour. It was found that 21 cm³ of acid was used.'
a Draw a labelled diagram of titration apparatus for this neutralization.
b What piece of apparatus should be used to accurately measure out 25 cm³ of sodium hydroxide solution?
c What colour was the solution in the flask at the start of the titration?
d What colour did it turn when the alkali had been neutralized?
e Was the acid more concentrated or less concentrated than the alkali? Explain your answer.
f Name the salt formed in this neutralization.
g Write an equation for the reaction.
h How would you obtain *pure* crystals of the salt?

6 A and B are white powders. A is insoluble in water but B is soluble and its solution has a pH of 3.
A **mixture** of A and B bubbles or effervesces in water. A gas is given off and a clear solution forms.
a One of the white powders is an acid. Is it A or B?
b The other white powder is a carbonate. What gas is given off in the reaction?
c Although A is insoluble in water, a clear solution forms when the mixture of A and B is added to water. Explain why.

7 The chemical name for aspirin is **2-ethanoyloxybenzoic acid.**
This acid is soluble in hot water.
a How would you expect an aqueous solution of aspirin to affect litmus paper?
b Do you think it is a strong acid or a weak one? Explain why you think so.
c What would you expect to see when baking soda is added to an aqueous solution of aspirin?

Method of preparation	Reactants	Salt formed	Other products
a acid + alkali	calcium hydroxide and nitric acid	calcium nitrate	water
b acid + metal	zinc and hydrochloric acid	....	
c acid + alkali	.... and potassium hydroxide	potassium sulphate	water only
d acid + carbonate	.... and	sodium chloride	water and ....
e acid + metal	.... and	iron(II) sulphate	....
f acid + ...	nitric acid and sodium hydroxide	....	....
g acid + base	.... and copper(II) oxide	copper(II) sulphate	....
h acid + ...	.... and	copper(II) sulphate	carbon dioxide and
i precipitation	silver nitrate and potassium chloride	....	....
j precipitation	lead nitrate and potassium iodide	....	....

8 The table above is about the preparation of salts. Copy it and fill in the missing details.

9 Pink cobalt chloride is a **hydrated** salt. That means it contains **water of crystallization**. Its formula is $COCl_2.6H_2O$.
Here are five other salts:

copper(II) sulphate $CuSO_4.5H_2O$ (aq)
sodium chloride $NaCl$
zinc chloride $ZnCl_2.6H_2O$
sodium carbonate $Na_2CO_3.10H_2O$
ammonium nitrate NH_4NO_3

a Which ones are hydrated?
b Which ones are anhydrous (i.e. contain no water of crystallization)?
c Which ones are coloured?
d Write down the formula of the compound known as **common salt**.
e Write down the formula of the salt which is often used as a fertilizer.
f Anhydrous copper(II) sulphate is a white powder. What is its formula?
g Hydrated copper(II) sulphate is blue. It turns white when it is heated. Explain why.
h i Anhydrous copper(II) sulphate could be used to test whether a liquid contains water. Explain how. How could you *prove* that the liquid was *pure* water?
ii When some zinc chloride crystals are heated gently, a solution of zinc chloride forms. Suggest a reason.

10 a Divide the following salts into *Soluble in water* and *Insoluble in water* (if you need help look at page 138):

sodium chloride   zinc chloride
calcium carbonate   sodium sulphate
potassium chloride   copper(II) sulphate
barium sulphate   lead sulphate
barium carbonate   lead nitrate
silver chloride   sodium carbonate
sodium citrate   ammonium carbonate

b Now write down two starting compounds that could be used to make each *insoluble* salt.

11 These diagrams show the stages in the preparation of copper(II) ethanoate, which is a salt of ethanoic acid.

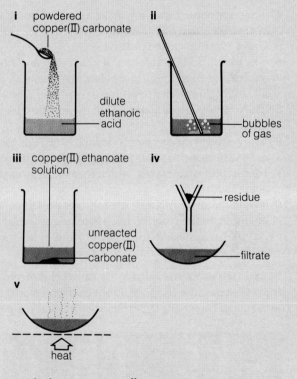

a Which gas is given off in stage **ii**?
b Write a word equation for the reaction in stage **ii**.
c How can you tell when the reaction is complete?
d Which reactant is completely used up in the reaction? Explain your answer.
e Why is copper(II) carbonate powder used, rather than lumps?
f Name the residue in stage **iv**.
g Write a list of instructions for carrying out this preparation in the laboratory.
h Suggest another copper compound that could be used instead of copper(II) carbonate, to make copper(II) ethanoate.

10.1 Metals and nonmetals

On page 26, you saw that there are 105 different elements. Of these, 84 are **metals** and 21 are **nonmetals**. Some of the metals are listed in the box on the right.

The properties of metals

Metals *usually* have these properties:

1 They are **strong** under tension and compression. That means they can withstand crushing and stretching without breaking.
2 They are **malleable**. That means they can be hammered and bent into shape without breaking.
3 They are **ductile**: they can be drawn out to make wires.
4 They are **sonorous**: they make a ringing noise when you strike them.
5 They are shiny when polished.
6 They are good conductors of electricity and heat.
7 They have high melting and boiling points. (They are all solid at room temperature, except mercury.)
8 They have high densities. That means they feel 'heavy'.
9 They react with oxygen to form oxides. For example, magnesium burns in air to form magnesium oxide. Metal oxides are **bases**, which means they react with acids to form salts.
10 When metals form ions, the ions are positive. For example, in the reaction between magnesium and oxygen, magnesium ions (Mg^{2+}) and oxide ions (O^{2-}) are formed, as shown on page 49.

The last two properties above are called **chemical properties**, because they are about chemical changes in the metals. The other properties are **physical properties**.

Some of the metals:

Aluminium, Al
Calcium, Ca
Copper, Cu
Gold, Au
Iron, Fe
Lead, Pb
Magnesium, Mg
Potassium, K
Silver, Ag
Sodium, Na
Tin, Sn
Zinc, Zn

What is density?

The density of a substance is a measure of how 'heavy' it is.

$$Density = \frac{mass\ (in\ grams)}{volume\ (in\ cm^3)}$$

Compare these:

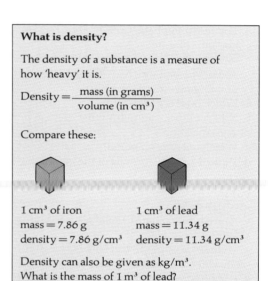

1 cm³ of iron
mass = 7.86 g
density = 7.86 g/cm³

1 cm³ of lead
mass = 11.34 g
density = 11.34 g/cm³

Density can also be given as kg/m³.
What is the mass of 1 m³ of lead?
What is its density?

Think of two reasons why metals are used to make drums . . .

. . . and then three reasons why they are used for saucepans.

All metals are different

The properties on the last page are typical of metals. But not all metals have *all* of these properties. For example:

Iron is a typical metal. It is used for gates like these because it is both malleable and strong. It is used for anchors because of its high density. It melts at 1530 °C. But unlike most other metals, it is **magnetic**.

Sodium is quite different. It is so soft that it can be cut with a knife, and it melts at only 98 °C. It is so light that it floats on water, but it reacts immediately with the water, forming a solution. No good for gates.

Gold melts at 1064 °C. Unlike most other metals it does not form an oxide – it is very unreactive. But it is malleable and ductile, and looks attractive. So it is used for making jewellery.

No two metals have exactly the same properties. You can find out more about the differences between them on the next four pages.

Comparing metals with nonmetals

Only 21 of the elements are nonmetals. Nonmetals are quite different from metals. They *usually* have these properties.

1 They are not strong, or malleable, or ductile, or sonorous. In fact, when solid nonmetals are hammered, they break up – they are **brittle**.
2 They have lower melting and boiling points than metals. (One of them is a liquid and eleven are gases, at room temperature.)
3 They are poor conductors of electricity. Graphite (carbon) is the only exception. They are also poor conductors of heat.
4 They have low densities.
5 Like metals, most of them react with oxygen to form oxides:

  sulphur + oxygen ⟶ sulphur dioxide

  But unlike metal oxides, these oxides are not bases. Many of them dissolve in water to give *acidic* solutions.
6 When they form ions, the ions are negative. Hydrogen is an exception – it forms the ion H^+.

Some of the nonmetals
Bromine, Br
Carbon, C
Chlorine, Cl
Helium, He
Hydrogen, H
Iodine, I
Nitrogen, N
Oxygen, O
Sulphur, S

Questions

1 Make two lists, showing twenty *metals* and fifteen *nonmetals*. Give their symbols too.
2 Try to think of a metal that is not malleable at room temperature.
3 Suggest reasons why:
  a silver is used for jewellery;
  b copper is used for electrical wiring.

4 For some uses, a highly sonorous metal is needed. Try to think of two examples.
5 Try to think of *two* reasons why:
  a mercury is used in thermometers;
  b aluminium is used for beer cans.
6 Look at the properties of nonmetals, above. Which are *physical* properties? Which are *chemical*?

10.2 Metals and reactivity (I)

On the last page you saw that all metals are different.
The next few pages compare the way some metals react, to see how different they are.

The reaction between metals and oxygen

Look at the way sodium reacts with oxygen:

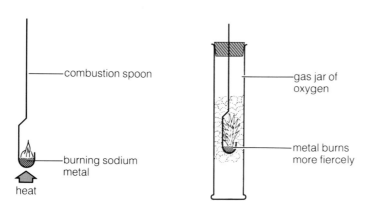

Because it reacts with oxygen, sodium is stored under oil.

A small piece of sodium is put in a combustion spoon and heated over a Bunsen flame. It melts quickly, and catches fire.

Then the spoon is plunged into a jar of oxygen. The metal burns even more fiercely, with a bright yellow flame.

The steps above can be repeated for other metals. This table shows what happens:

Metal	Behaviour	Order of reactivity	Product
Sodium	Catches fire with only a little heating. Burns fiercely with a bright yellow flame	most reactive	Sodium peroxide, Na_2O_2, a pale yellow powder
Magnesium	Catches fire easily. Burns with a blinding white flame		Magnesium oxide, MgO, a white powder
Iron	Does not burn, but the hot metal glows brightly in oxygen, and gives off yellow sparks		Iron oxide, Fe_3O_4, a black powder
Copper	Does not burn, but the hot metal becomes coated with a black substance		Copper oxide, CuO, a black powder
Gold	No reaction, no matter how much the metal is heated	least reactive	——

If a reaction takes place, the product is an oxide.
Sodium reacts the most vigorously with oxygen. It is the **most reactive** of the five metals. Gold does not react at all — it is the least reactive of them. The arrow on the table shows the **order of reactivity**.

The reaction between metals and water

Metals also show differences in the way they react with water. For example:

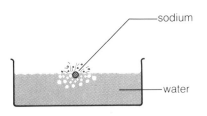

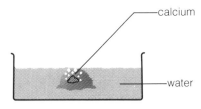

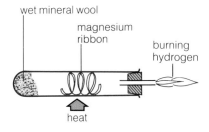

Sodium reacts violently with cold water, whizzing over the surface. Hydrogen gas and a clear solution of sodium hydroxide are formed.

The reaction between calcium and cold water is slower. Hydrogen bubbles off, and a cloudy solution of calcium hydroxide forms.

Magnesium reacts very slowly with cold water, but vigorously when heated in steam: it glows brightly. Hydrogen and solid magnesium oxide are formed.

This table shows the results for other metals too:

Metal	Reaction	Order of reactivity	Products
Potassium	Very violent with cold water. Catches fire	most reactive	Hydrogen and a solution of potassium hydroxide, KOH
Sodium	Violent with cold water		Hydrogen and a solution of sodium hydroxide, NaOH
Calcium	Less violent with cold water		Hydrogen and calcium hydroxide, $Ca(OH)_2$, which is only slightly soluble
Magnesium	Very slow with cold water, but vigorous with steam		Hydrogen and solid magnesium oxide, MgO
Zinc	Quite slow with steam		Hydrogen and solid zinc oxide, ZnO
Iron	Slow with steam		Hydrogen and solid iron oxide, Fe_3O_4
Copper Gold	No reaction	least reactive	——

Notice that the first three metals in the list produce hydroxides. The others produce oxides, if they react at all.

Now compare this table with the one on the opposite page. Is sodium more reactive than iron each time? Is iron more reactive than copper each time?

Questions

1 Describe how magnesium and iron each react with oxygen. Write balanced equations for the reactions.
2 Which is more reactive, copper or iron?
3 Which is more reactive, sodium or zinc?

4 What gas is always produced if a metal reacts with water?
5 Describe how magnesium reacts with steam. Write a balanced equation for the reaction.

10.3 Metals and reactivity (II)

Reaction with hydrochloric acid

On the last two pages, you saw that different metals react differently with oxygen and water. They also react differently with **acids**. Compare these results using hydrochloric acid:

Metal	Reaction with hydrochloric acid	Order of reactivity	Products
Magnesium	Vigorous	most reactive	Hydrogen and a solution of magnesium chloride, $MgCl_2$
Zinc	Quite slow		Hydrogen and a solution of zinc chloride, $ZnCl_2$
Iron	Slow		Hydrogen and a solution of iron(II) chloride, $FeCl_2$
Lead	Slow, and only if the acid is concentrated		Hydrogen and a solution of lead(II) chloride, $PbCl_2$
Copper Gold	No reaction, even with concentrated acid	least reactive	

Now compare this table with the last two tables. Is iron always more reactive than copper? Is magnesium always more reactive than iron?

Competing for oxygen

The reactions with oxygen, water, and hydrochloric acid show that iron is more reactive than copper. Now look at this experiment.

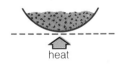

This is a mixture of powdered iron and copper(II) oxide. On heating, the reaction starts.

The mixture glows, even after the Bunsen is removed. Iron(II) oxide and copper are formed.

Here iron and copper are competing for oxygen. Iron wins:

iron + copper(II) oxide ⟶ iron(II) oxide + copper
$Fe\,(s)$ + $CuO\,(s)$ ⟶ $FeO\,(s)$ + $Cu\,(s)$

By taking away the oxygen from copper, iron is acting as a **reducing agent** (page 86). Other metals behave in the same way when heated with the oxides of less reactive metals.
When a metal is heated with the oxide of a less reactive metal, it will remove the oxygen from it. The reaction is exothermic.

Using a competition reaction to repair railway lines. Aluminium and iron(III) oxide are being heated together, to give molten iron. This is run into gaps between rails. The process is called the Thermit process.

Displacement reactions

Let's look at another reaction involving iron and copper.

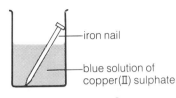

This time, an iron nail is placed in some blue copper(II) sulphate solution.

Soon a coat of copper appears on the nail. The solution turns pale green.

Here iron and copper are competing to be the compound in solution. Once again iron wins. It drives out or **displaces** copper from the copper(II) sulphate solution. Green iron(II) sulphate is formed:

iron + copper(II) sulphate $\longrightarrow$ iron(II) sulphate + copper

$$Fe\ (s) + \underset{\text{blue}}{CuSO_4\ (aq)} \longrightarrow \underset{\text{green}}{FeSO_4\ (aq)} + Cu\ (s)$$

Other metals displace less reactive metals in the same way. **A metal will always displace a less reactive metal from solutions of its compounds.**

Now look at the way copper reacts with silver nitrate solution:

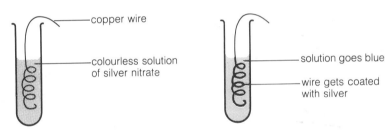

A coil of copper is placed in a solution of silver nitrate. The copper gradually dissolves ...

... so the solution goes blue. At the same time, a coating of silver forms on the wire.

Which metal is more reactive, copper or silver? (Check the reactivity series on the next page to see if you were right.)

Crystals of silver on a copper wire. The copper has displaced the silver from silver nitrate solution.

Questions

The table on the opposite page will help you answer some of these questions.

1 Describe how iron and lead react with hydrochloric acid. Write balanced equations for the reactions.
2 Write a rule for the reaction of a metal with:
  a oxides of other metals;
  b solutions of compounds of other metals.
3 Will copper react with lead(II) oxide? Explain why.

4 Will iron react with lead(II) oxide? If *yes*, write a balanced equation for the reaction.
5 What would you *see* when zinc is added to copper(II) sulphate solution? (Zinc sulphate is colourless.)
6 Explain how the Thermit process works.
7 Tin does not react with iron(II) oxide. But it reduces lead(II) oxide to lead. Arrange tin, iron and lead in order of decreasing reactivity.

10.4 The reactivity series

By comparing the reactions of metals with oxygen, water, acid, metal oxides, and solutions of metal salts, we can arrange the metals in order of reactivity. The list is called the **reactivity series**. Here it is:

Potassium, K — most reactive
Sodium, Na
Calcium, Ca
Magnesium, Mg
Aluminium, Al
Zinc, Zn — increasing reactivity
Iron, Fe
Lead, Pb

metals above this line react with acids, displacing **hydrogen**

Copper, Cu
Silver, Ag
Gold, Au — least reactive

A metal's position in the reactivity series gives us a clue about its possible uses. Only unreactive metals are used to make coins.

Things to remember about the reactivity series

1 The more reactive the metal, the more it 'likes' to form compounds. So only copper, silver and gold are ever found as *elements* in the Earth's crust. The other metals are *always* found as compounds.

2 The more reactive the metal, the more **stable** its compounds. A stable compound is difficult to break down or decompose because the bonds holding it together are very strong.

3 So the more reactive the metal, the more difficult it is to **extract** from its compounds. For the most reactive metals you'll need the toughest method of extraction: electrolysis. Sodium is extracted from sodium chloride using electrolysis.

4 When a metal reacts, it gives up electrons to form ions. **The more reactive the metal, the more easily it gives up electrons.**

5 Many metals occur naturally as oxides. You could extract them by heating the oxide with a more reactive metal, which will steal the oxygen away. For example, you could use aluminium to extract iron from iron oxide. (But it's much cheaper to use carbon monoxide — see page 152.)

6 If you put two metals together in air, touching, the more reactive one will corrode first. The other won't start until the first has corroded away. So zinc is often used to stop iron rusting — see page 161.

7 Since copper, silver and gold occur as *elements* in the Earth, they are easy to extract. So humans have been using them for thousands of years.
   But aluminium didn't get widely used until around 1890, not long after electrolysis was invented.
   The dates on the right show when different metals began to be widely used. Compare the list with the reactivity series. What do you notice?

When metals began to be widely used …	
Aluminium	1890
Zinc	1500
Iron	1400 BC
Lead	2000 BC
Copper	8000 BC

The reactivity series and cells

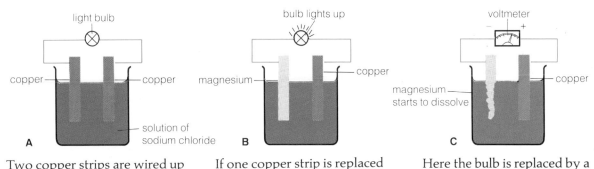

A

Two copper strips are wired up to a light bulb and placed in a solution of sodium chloride. The solution is an **electrolyte** — it *can* conduct electricity. But nothing happens.

B

If one copper strip is replaced by a *magnesium* strip, the bulb lights up. Electricity is being produced. Electrons are flowing through the wires even though there is no battery.

C

Here the bulb is replaced by a **voltmeter**. This measures the 'push' or **voltage** that makes the electrons flow. It is 2.7 volts. The needle shows the direction of the electron flow.

Where do the electrons come from? The answer is this:
Magnesium can give up electrons more readily than copper. So magnesium atoms give up electrons and go into solution as ions. The electrons flow along the wire to the copper strip.
This arrangement is called a **cell**. The magnesium strip is the **negative pole** of the cell. The copper strip is the **positive pole**.

A cell consists of two different metals and an electrolyte. In the cell, chemical energy produces electricity. The more reactive metal becomes the negative pole from which electrons flow.

The experiment can be repeated with other metals. As long as the strips are made of different metals, electrons will flow. But the voltage changes with the metals, as this table shows.

Metal strips	Volts
Copper and magnesium	2.70
Copper and iron	0.78
Lead and zinc	0.64
Lead and iron	0.32

Of these metals, copper and magnesium are furthest apart in the reactivity series. They give the highest voltage. Lead and iron are closest. Look at the voltage they give.
The further apart the metals are in the reactivity series, the higher the voltage of the cell.

The correct name for a torch battery is a 'dry cell'. It contains two different metals and an electrolyte. The electrolyte is a paste rather than a liquid, because a liquid would leak.

Questions

1 Why is sodium never found uncombined in nature?
2 Which will break down more easily on heating, magnesium nitrate or silver nitrate? Explain.
3 Why is magnesium more reactive than copper?
4 Explain why the bulb lights in experiment B above.
5 Why doesn't the bulb light in experiment A?

6 Will the bulb in B light if a sugar solution is used instead? Explain your answer.
7 Which pair of metals gives the highest voltage? Iron, zinc; copper, iron; silver, magnesium. Why?
8 In question 7, which metal in each pair becomes the negative pole?

10.5 Metals in the Earth's crust

We get some metals from the sea, but most from the Earth's **crust**. The Earth's crust is the outer layer of the Earth. It is only about 6 km thick under the oceans, and about 70 km thick under the highest mountain. But no one has yet managed to dig a hole through it to reach the next layer!

The composition of the Earth's crust

The Earth's crust is a huge mixture of many different compounds. It also contains *elements* such as sulphur, copper, silver, platinum and gold. These occur uncombined, or **native**, because they are unreactive. If you could dig up the Earth's crust and break down all the compounds, you would find it is almost half oxygen! Its composition is:

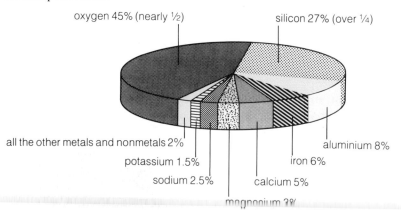

oxygen 45% (nearly ½)   silicon 27% (over ¼)

all the other metals and nonmetals 2%
potassium 1.5%
sodium 2.5%
magnesium 2%
calcium 5%
iron 6%
aluminium 8%

Iron is a very important metal. We use about nine times more iron than all the other metals put together. It is made into steel and used for large things like the Forth bridge (above) as well as small things like needles.

Nearly three-quarters of the Earth's crust is made of just two nonmetals, oxygen and silicon. These are found combined in compounds such as silicon dioxide (**silica** or **sand**). Oxygen also occurs in compounds such as aluminium oxide, iron oxide and calcium carbonate.

The rest of the crust consists mainly of just six metals. Aluminium is the most abundant of these, and iron is next. None of the six is found uncombined, because they are all reactive metals, and reactive metals 'like' to form compounds.

Some metals are scarce

Look again at the pie chart above. It shows that there are just six plentiful metals. All the other metals *together* make up less than one-fiftieth of the Earth's crust, so they are not plentiful.

If a metal makes up less than one-thousandth of the Earth's crust, it is called **scarce**. Here are some of the scarce metals:

copper	zinc	lead	tin
mercury	silver	gold	platinum

The scarce metals are expensive to buy. Even so, we are using them up very quickly, and many people are worried that they will soon run out.

A wedding kimono woven from platinum, a scarce metal. It is worth £70 000.

Metal ores

The rocks in the Earth's crust are a mixture of many different compounds. But some contain a large amount of a single metal compound, or a single metal, and it may be worth digging these up to extract the metal. Rocks from which metals are obtained are called **ores**. Below are some examples.

This is a chunk of **rock salt**, the main ore of sodium. It is mostly **sodium chloride**. Nearly all countries have some rock salt. The largest deposits are in Russia and the USA.

This is a piece of **bauxite**, the main ore of aluminium. It is mostly **aluminium oxide**. Huge amounts of it are found in Australia and Jamaica. But there is none in Britain.

Since gold is unreactive, it is found as a free element. This picture shows a lump of almost pure gold. Most of the world's gold comes from South Africa.

To mine or not to mine?

Before starting to mine an ore, the mining company must decide whether it is economical. It must find the answers to questions like these:

1 How much ore is there?
2 How much metal will we get from it?
3 Are there any special problems about getting the ore out?
4 How much will it cost to mine the ore and extract the metal from it? (The cost will include roads, buildings, mining equipment, extraction plant, transport, fuel, chemicals, and wages.)
5 How much will we be able to sell the metal for?
6 So will we make a profit if we go ahead?

The answers to these questions will change from year to year. If the cost of fuel drops, and the selling price of a metal rises, even a low-quality or **low-grade** ore may become worth mining.

The mining company must also think about the local community, who may worry that the landscape will be ruined by dust, pits and scars, and that the extraction plant will cause pollution. On the other hand, they may welcome the new jobs that mining will bring.

A landscape spoiled by mining. These days, mining companies are often forced to clear and restore the land when a mine is exhausted. That is another cost to consider.

Questions

1 Which is the main *element* in the Earth's crust?
2 Which is the most common *metal* in the Earth's crust? Which is the second?
3 Gold occurs *native* in the Earth's crust. Explain.
4 Is it true that the reactive metals are plentiful in the Earth's crust?

5 What is a *scarce* metal? Name four.
6 One metal is used more than all the others put together. Which one?
7 What is a metal ore?
8 Name the main ore of: **a** sodium **b** aluminium What is the main compound in each ore?

10.6 Extracting metals from their ores

Ways of extracting metals

On the last page you saw that metals are obtained from **ores**. An ore is usually a compound of the metal, mixed with impurities. When the ore has been dug up, it must be decomposed in some way to get the metal. This is called **extracting** the metal.

The method of extraction depends on how reactive the metal is. **The more reactive the metal, the more difficult its compounds are to decompose.**

Electrolysis is the most powerful way to decompose a metal compound. But it needs a lot of electricity, and that makes it expensive. So it is used only for the more reactive metals, as this table shows.

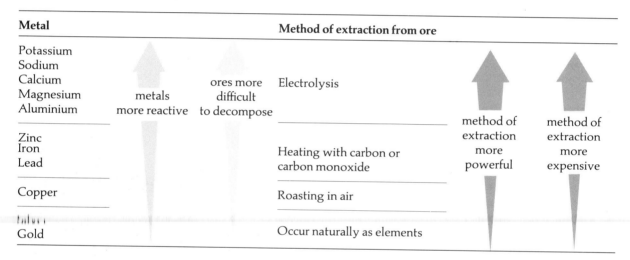

Metal			Method of extraction from ore		
Potassium Sodium Calcium Magnesium Aluminium	metals more reactive	ores more difficult to decompose	Electrolysis	method of extraction more powerful	method of extraction more expensive
Zinc Iron Lead			Heating with carbon or carbon monoxide		
Copper			Roasting in air		
Silver Gold			Occur naturally as elements		

Examples of the different methods of extraction

1 **Electrolysis.** This is used for extracting sodium from rock salt. The rock salt is first melted in giant steel tanks:

sodium chloride $\longrightarrow$ sodium + chlorine
$2NaCl\,(l)$ $\longrightarrow$ $2Na\,(l)$ + $Cl_2\,(g)$

Electrolysis is also used for extracting aluminium (page 156).

2 **Heating with carbon monoxide.** This is used for extracting iron from iron ore in the blast furnace (page 158):

iron(III) oxide + carbon monoxide $\longrightarrow$ iron + carbon dioxide
$Fe_2O_3\,(s)$ + $3CO\,(g)$ $\longrightarrow$ $2Fe\,(l)$ + $3CO_2\,(g)$

3 **Roasting in air.** Some copper is found native. But most occurs as copper(I) sulphide, in an ore called **copper pyrites**. The copper is extracted by roasting the sulphide in air:

copper(I) sulphide + oxygen $\longrightarrow$ copper + sulphur dioxide
$Cu_2S\,(s)$ + $O_2\,(g)$ $\longrightarrow$ $2Cu\,(l)$ + $SO_2\,(g)$

Until a hundred years ago, aluminium was rarely used because it was difficult to extract. Then in 1886 the modern process of electrolysis was developed. The statue of Eros in London's Piccadilly Circus was made from aluminium in 1893.

The recycling of metals

Metals are a **non-renewable** resource. In other words, when you dig up an ore, no new ore forms to replace it. So the supply will eventually run out. Look at this table:

Metal	Amount used up each year (million tonnes)	Number of years before the metal runs out (approx)
Iron	800	110
Aluminium	12	350
Copper	8	40
Zinc	4.5	60
Lead	4	20
Tin	0.25	15

The electrolysis of aluminium is very expensive, because of all the electricity it uses. Scrap aluminium can be melted down for only a twentieth of that cost.

These are only rough figures. But they show how careful we need to be. It makes sense to **recycle** metals – that is, to melt down used metals, rather than throw them away.

A metal company will recycle metal only if it is economical. The company has to work out the cost of collecting the scrap, transporting it, melting it down, getting rid of impurities, and paying the workers' wages. In future, as metals get scarcer and more expensive, recycling will become a more important process.

Scrap iron can be separated from other metals using a magnet.

Every year people in Britain use and throw away 10 000 million tins. The metal in them is worth millions of pounds. But unfortunately it is not easy to separate the layer of tin from the steel can.

Questions

1 What does the *extraction* of a metal mean?
2 For the extraction of sodium, name:
   a the ore used;   b the method used.
   Why must the ore be melted first?
3 Aluminium is more common than iron in the Earth's crust, but more *expensive* too. Suggest why.
4 Why is electrolysis needed for extracting calcium? Why is it *not* needed for extracting copper?
5 Metals are a non-renewable resource. It makes sense to recycle them. Explain the underlined words.
6 Which property of iron can be used to separate it from other kinds of scrap metal?

10.7 Making use of metals

Pure metals and alloys

The way a metal is used depends on its **properties**:

Pure aluminium can be rolled into very thin sheets, which are quite strong but easily cut. So it is used for milk bottle tops and cooking foil.

Pure lead is soft, and bends easily without being heated. It also resists corrosion. So it is used to seal off brickwork around chimneys.

Pure copper is easily drawn into wires, and is an excellent conductor of electricity. So it is used for electrical wiring around the home.

Sometimes a metal is most useful when it is pure. For example, copper is not nearly such a good conductor when it contains impurities.

But many metals are more useful when they are *not* pure. Iron is the most widely-used metal of all, and it is almost never used pure:

Pure iron is no good for building things, because it is too soft and stretches easily, as you can see in the photo above. Besides, it rusts easily too.

But when a little carbon (0.5%) is mixed with it, the result is **mild steel**. This is hard and strong. It is used for buildings, bridges, ships and car bodies.

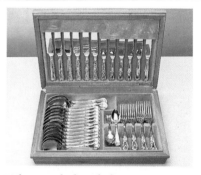

When nickel and chromium are mixed with iron, the result is **stainless steel**. This is hard and rustproof. It is used for car parts, kitchen sinks, and cutlery.

You can see that the properties of the iron have been changed by mixing other substances with it.
The properties of any metal can be changed by mixing other substances with it. The mixtures are called alloys.
The added substances are usually metals, but sometimes nonmetals like carbon or silicon. An alloy is usually made by melting the main metal and then dissolving the other substances in it.

Uses of pure metals

This table summarizes some uses of pure metals:

Metal	Uses	Properties that make it suitable
Sodium	A coolant in nuclear reactors	Conducts heat well. Melts at only 98 °C, so the hot metal will flow along pipes
Aluminium	Overhead electricity cables	A good conductor of electricity (not as good as copper, but cheaper and much lighter).
	Saucepans	Conducts heat well and is non-toxic (non-poisonous)
Zinc	Coating iron, to give **galvanized** iron	Protects the iron from rusting
Tin	Coating steel cans or 'tins'	Unreactive and non-toxic. Protects the steel from rusting
Mercury	Thermometers	Liquid at room temperature. Expands on heating. Easy to see, and does not wet the sides of tubes

Notice that sodium is not used for making things — it is too reactive.
But in nuclear reactors it is well protected from any reactions.

Uses of alloys

There are thousands of different alloys. Here are just a few!

Alloy	Made from	Special properties	Uses
Cupronickel	75% copper 25% nickel	Hard-wearing, attractive silver colour	'Silver' coins
Stainless steel	70% iron 20% chromium 10% nickel	Does not rust	Car parts, kitchen sinks, cutlery
Duralium	96% aluminium 4% copper	Light, and stronger than aluminium	Aircraft parts
Brass	70% copper 30% zinc	Harder than copper, does not corrode	Musical instruments
Bronze	95% copper 5% tin	Harder than brass, does not corrode, sonorous	Statues, ornaments, church bells
Solder	70% tin 30% lead	Low melting point	Joining wires and pipes

You can find out more about the alloys of iron on page 159.

Questions

1 Why is iron more useful when it is mixed with a little carbon?
2 What are alloys? How are they made?
3 Explain why tin is used to coat food tins.
4 Name an alloy that:
  a has a low melting point;   b never rusts.
5 Which metals are used to make:
  stainless steel?   duralium?   bronze?

10.8 More about aluminium

From rocks to rockets

Aluminium is the most abundant metal in the Earth's crust. Its main ore is **bauxite**, which is aluminium oxide mixed with impurities like sand and iron oxide. The impurities make it reddish brown.

These are the steps in obtaining aluminium:

1 First, geologists test rocks to find out how much bauxite there is, and whether it is worth mining. If the tests are satisfactory, mining begins.

2 Bauxite usually lies near the surface, so it is easy to dig up. This is a bauxite mine in Jamaica. Everything gets coated with red-brown bauxite dust.

3 From the mine, the ore is taken to a bauxite plant, where it is treated to remove the impurities. The result is white **aluminium oxide**, or **alumina**.

4 The alumina is taken to another plant for electrolysis. Much of the Jamaican alumina is shipped to plants like this one, in Canada or the USA . . .

5 . . . where electricity is cheaper. There it is electrolysed to give aluminium. The metal is made into sheets and blocks, and sold to other industries.

6 It is used to make beer cans, cooking foil, saucepans, racing bikes, TV aerials, electricity cables, ships, aeroplanes and even space rockets.

A closer look at the electrolysis

In step 5 above you saw that aluminium is obtained from alumina by electrolysis.
The electrolysis is carried out in a huge steel tank. The tank is lined with graphite, which acts as the cathode. Huge blocks of graphite hang in the middle of the tank, and act as anodes.

Pure alumina melts at 2045 °C. It would be expensive, and dangerous, to keep the tank at that temperature. Instead, the alumina is dissolved in molten **cryolite** for the electrolysis. (Cryolite is another aluminium compound, with a much lower melting point.)

The steel tanks in which electrolysis takes place.

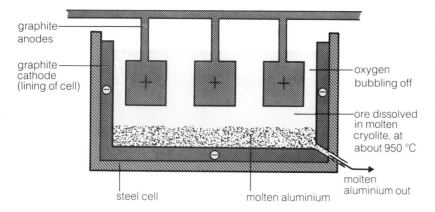

When the alumina dissolves, its aluminium ions and oxide ions become free to move.

At the cathode   The aluminium ions receive electrons:

$$4Al^{3+} + 12e^- \longrightarrow 4Al$$

The aluminium atoms collect together, and drop to the bottom of the cell as molten metal. This is run off at intervals.

At the anodes   The oxygen ions give up electrons:

$$6O^{2-} \longrightarrow 3O_2 + 12e^-$$

Oxygen gas bubbles off. But unfortunately it attacks the graphite anodes and eats them away, so they must be replaced from time to time.

Some properties of aluminium

1 Aluminium is a bluish-silver, shiny metal.
2 Unlike most metals, it has a low density – it is 'light'.
3 It is a good conductor of heat and electricity.
4 It is malleable and ductile.
5 It is non-toxic.
6 It is not very strong when pure, but it can be made stronger by mixing it with other metals to form alloys (page 154).

These properties lead to the wide range of uses for aluminium, given in step 6 on the opposite page.

This underground train is made of aluminium strengthened with small amounts of other metals.

Questions

1 Copy and complete: The chief ore of aluminium is called .... It is first purified to give .... or ...., which has the formula .... Then this is .... to give aluminium.
2 Draw the cell for the electrolysis of aluminium.

3 Why do the aluminium ions move to the cathode?
4 What happens at the cathode?
5 Why must the anodes be replaced from time to time?
6 List six uses of aluminium. For each, say what properties of the metal make it suitable.

10.9 More about iron

The extraction of iron

Iron is the second most abundant metal in the Earth's crust. To extract it, three substances are needed:

1 **Iron ore**. The chief ore of iron is called **haematite**. It is mainly iron oxide, Fe_2O_3, mixed with sand.
2 **Limestone**. This is a common rock. It is mainly calcium carbonate, $CaCO_3$.
3 **Coke**. This is made from coal, and is almost pure carbon.

These three substances are mixed together to give a mixture called **charge**. The charge is heated in a tall oven called a **blast furnace**. Several reactions take place, and finally liquid iron is produced.

In the blast furnace   A blast furnace is like a giant chimney, at least 30 m tall. It is made of steel, and lined with fireproof bricks. The charge is added through the top. Hot air is blasted through the bottom, making the charge glow white-hot. These reactions take place:

1 The coke reacts with oxygen in the air, giving **carbon dioxide**:

$$C\,(s)\ +\ O_2\,(g)\ \longrightarrow\ CO_2\,(g)$$

2 The limestone decomposes to **calcium oxide** and **carbon dioxide**:

3 The carbon dioxide reacts with more coke, giving **carbon monoxide**:

$$C\,(s)\ +\ CO_2\,(g)\ \longrightarrow\ 2CO\,(g)$$

4 This reacts with iron oxide in the ore, giving liquid **iron**:

$$Fe_2O_3\,(s)\ +\ 3CO\,(g)\ \longrightarrow\ 2Fe\,(l)\ +\ 3CO_2\,(g)$$

The iron trickles to the bottom of the furnace.

5 Calcium oxide from step 2 reacts with sand in the ore, to form **calcium silicate** or **slag**:

$$CaO\,(s)\ +\ SiO_2\,(s)\ \longrightarrow\ CaSiO_3\,(l)$$

The slag runs down the furnace and floats on the iron.
The slag and iron are drained from the bottom of the furnace. When the slag solidifies it is sold, mostly for road-building. Only *some* of the iron is left to solidify, in moulds. The rest is taken away while still hot, and turned into steel.

A stockpile of iron ore.

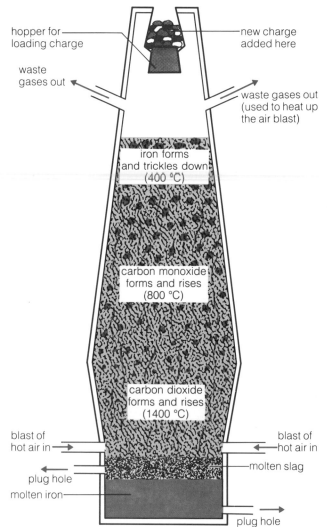

hopper for loading charge

new charge added here

waste gases out

waste gases out (used to heat up the air blast)

iron forms and trickles down (400 °C)

carbon monoxide forms and rises (800 °C)

carbon dioxide forms and rises (1400 °C)

blast of hot air in

blast of hot air in

molten slag

plug hole

molten iron

plug hole

Cast iron

The iron from the blast furnace usually contains a lot of carbon (up to 4%) as well as other impurities. You saw that some of the iron is allowed to solidify in moulds or **casts**. It is called **cast iron**. The carbon makes it very hard, but also brittle – it snaps under strain. So these days is only used for things like gas cylinders, railings and storage tanks, which are not likely to get bent during use.

The famous bridge at Ironbridge is made of cast iron. Opened in 1781, it was the first iron bridge in the world. Modern bridges are made of steel.

Steel

Most of the iron from the blast furnace is turned into **steel**.
Steel is an alloy of iron. There are many different types of steel, and each has different properties. This is how they are made:

1 **First, unwanted impurities are removed from the iron.** This is done in an **oxygen furnace**. The molten metal is poured into the furnace, as shown on the right. A lot of scrap iron is added too. Then some calcium oxide is added, and a jet of oxygen is turned on. The calcium oxide reacts with some of the impurities, forming a slag that can be skimmed off. Oxygen reacts with the others, and they burn away.

   For some steels, *all* the impurities are removed from the iron. But many steels are just iron plus a small amount of carbon – enough to make the metal hard, but not brittle. So the carbon content has to be checked continually. When it is correct, the oxygen is turned off.

2 **Then, other elements may be added.** As you saw on page 154, different elements affect iron in different ways. The added elements are carefully measured out, to give steels of exactly the required properties.

Molten iron being poured into an oxygen furnace.

There are thousands of different steels. Below are just three of them:

Name	Contains			Special property	Uses
Mild steel	99.5% Fe,	0.5% C		Hard but easily worked	Buildings, car bodies
Hard steel	99% Fe,	1% C		Very hard	Blades for cutting tools
Duriron	84% Fe,	1% C,	15% Si	Not affected by acid	Tanks and pipes in chemical factories

Questions

1 Name the raw materials for extracting iron.
2 Write an equation for the reaction that gives iron.
3 The calcium carbonate in the blast furnace helps to purify the iron. Explain how, with equations.
4 Name a waste gas from the furnace.

5 The slag and waste gas are both useful. How?
6 What is cast iron? Why is it brittle?
7 Explain how mild steel is made.
8 What makes *hard steel* harder than mild steel?
9 Explain why most iron is turned into steel.

10.10 Corrosion

When a metal is attacked by air, water or other substances in its surroundings, the metal is said to **corrode**:

Damp air quickly attacks potassium, turning it into a pool of potassium hydroxide solution. (That is why potassium is stored under oil!)

Iron also corrodes in damp air. So do most steels. The result is **rust** and the process is called **rusting**. It happens especially quickly near sea water.

But gold is unreactive and never corrodes. This gold mask of King Tutankhamun was buried in his tomb for over 3000 years, and still looks as good as new.

In general, the more reactive a metal is, the more readily it corrodes.

The corrosion of iron and steel

The corrosion of iron and steel is called **rusting**. The iron is **oxidized**. It needs both air and water, as these tests show.

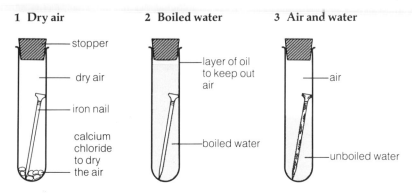

1 Dry air
- stopper
- dry air
- iron nail
- calcium chloride to dry the air

2 Boiled water
- layer of oil to keep out air
- boiled water

3 Air and water
- air
- unboiled water

Nails 1 and 2 do not rust. Nail 3 does rust.

How to stop rust.   When iron is made into stainless steel, it does *not* rust. But stainless steel is too expensive to use in large amounts. So other methods of rust prevention are needed.
Below are some of the methods. They mostly involve *coating the metal with something, to keep out air and water*:

1 **Paint.** Steel bridges and railings are usually painted. Paints that contain lead or zinc are mostly used, because these are especially good at preventing rust. For example, 'red lead' paints contain an oxide of lead, Pb_3O_4.
2 **Grease.** Tools, and machine parts, are coated with grease or oil.

Steel girders for building bridges are first sprayed with special paint.

3 **Plastic.** Steel is coated with plastic for use in garden chairs, bicycle baskets and dish racks. Plastic is cheap and can be made to look attractive.

4 **Galvanizing.** Iron for sheds and dustbins is usually coated with zinc. It is called **galvanized iron**.

5 **Tin plating.** Baked beans come in 'tins' which are made from steel coated on both sides with a fine layer of tin. Tin is used because it is unreactive, and non-toxic. It is deposited on the steel by electrolysis, in a process called **tin plating**.

6 **Chromium plating.** Chromium is used to coat steel with a shiny protective layer, for example on car bumpers. Like tin, the chromium is deposited by electrolysis.

7 **Sacrificial protection.** Magnesium is more reactive than iron. When a bar of magnesium is attached to the side of a steel ship, it corrodes instead of the steel. When it is nearly eaten away it can be replaced by a fresh bar. This is called **sacrificial protection**, because the magnesium is sacrificed to protect the steel. Zinc can be used in the same way.

Steel cans for food are plated with tin.

Does aluminium corrode?

Aluminium is more reactive than iron, so you might expect it to corrode faster in damp air. In fact, clean aluminium starts corroding immediately, but the reaction quickly stops. These diagrams show why.

rust flakes

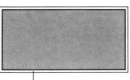

coat of aluminium oxide

When iron corrodes, rust forms in tiny flakes. Damp air can get past the flakes, to attack the metal below. In time, it rusts all the way through.

But when aluminium reacts with air, an even coat of aluminium oxide forms. This seals the metal surface and protects it from further attack.

The layer of aluminium oxide can be made thicker by electrolysis, to give even more protection. This process is called **anodizing**. The aluminium is used as the anode of a cell in which dilute sulphuric acid is electrolysed. Oxygen forms at the anode and reacts with the aluminium, so the layer of oxide grows.

Anodized aluminium is used for cookers, fridges, cooking utensils, saucepans, window frames and sometimes for wall panels on buildings. The oxide layer can easily be dyed to a bright colour.

Anodized aluminium is used for door and window frames.

Questions

1 What is *corrosion*?
2 What two substances cause rusting?
3 Iron that is tin-plated does not rust. Why?
4 In one method of rust prevention, the iron is not coated with anything. Which method is this?

5 Explain why magnesium can prevent the rusting of iron. Why would zinc do instead?
6 Why doesn't aluminium corrode right through?
7 How can a layer of aluminium oxide be made thicker? What is the process called?

Questions on Section 10

1 Read the following passage about the physical properties of metals.
Elements are divided into metals and nonmetals. All metals are <u>electrical conductors</u>. Many of them have a high <u>density</u> and they are usually <u>ductile</u> and <u>malleable</u>. All these properties influence the way the metals are used. Some metals are <u>sonorous</u> and this leads to special uses for them.
a Explain the meaning of the words underlined.
b Copper is ductile. How is this property useful in everyday life?
c Aluminium is hammered and bent to make large structures for use in ships and aeroplanes. What property is important in the shaping of this metal?
d Name one metal that has a *low* density.
e Some metals are cast into bells. What property must the chosen metals have?
f Add the correct word: *Metals are good conductors of* .... *and electricity.*
g Name one other physical property of metals and give two examples of how this property is useful.

2 **a** Write a short passage, like that in question 1, about the physical properties of nonmetals.
b Give one way in which the *chemical* properties of nonmetals and metals differ.

3 This is a list of metals in order of their chemical reactivity:

> *sodium (most reactive)*
> *calcium*
> *magnesium*
> *zinc*
> *iron*
> *lead*
> *copper*
> *silver (least reactive)*

a Which element is stored in oil?
b Which elements will react with *cold* water?
c Choose one metal that will not react with cold water but will react with steam. Draw a diagram of suitable apparatus to demonstrate this reaction. (You must show how the steam is produced.)
d Name the gas given off in **b** and **c**.
e Name another reagent that reacts with metals to give the same gas.
f Which of the metals will *not* react with oxygen when heated?
g How does iron react when heated in oxygen?
h How would you expect:
   i lead   **ii** calcium
to react when heated in oxygen?
(Hint: look at the table on page 144 and the full reactivity series on page 148.)

4 Look again at the list of metals in question 3. Because zinc is more reactive than iron, it will remove the oxygen from iron(III) oxide, on heating.
a Write a word equation for the reaction.
b Decide whether these chemicals will react together, when heated:
   i magnesium + lead(II) oxide
   ii copper + lead(II) oxide
   iii magnesium + copper(II) oxide
c For those which react:
   i describe what you would *see*;
   ii write a word equation.
d What name is given to reactions of this type?

5 When magnesium powder is added to copper(II) sulphate solution, solid copper forms. This change occurs because magnesium is more reactive than copper.
a Write a word equation for the reaction.
b Use the list of metals in question 3 to decide whether these will react together:
   i iron + copper(II) sulphate solution
   ii silver + calcium nitrate solution
   iii zinc + lead(II) nitrate solution
c For those which react:
   i describe what you would *see*;
   ii write the word equation.
d What name is given to reactions of this type?

6 Choose one metal to fit each description below. (You must choose a different one each time.) Write down the name of the metal. Then write a balanced equation for the reaction that takes place.
a A metal which burns in oxygen.
b A metal which reacts with oxygen without burning.
c A metal which reacts gently with dilute hydrochloric acid.
d A metal which floats on water and reacts vigorously with it.

7 Strips of copper foil and magnesium ribbon were cleaned with sandpaper and then connected as shown below. The bulb lit up.

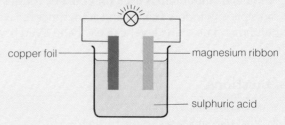

a Why were the metals first cleaned with sandpaper?
b Name the electrolyte used.

c Explain why the bulb lit up.

d Which is the more reactive of the two metals?

e Which of the two metals releases electrons into the circuit?

f What could replace the bulb to measure the 'push' that makes the electrons flow?

g What energy change takes place while the bulb is lit?

h What name is usually given to this type of arrangement?

i Give two good reasons why this particular arrangement could not be used commercially as a battery.

8 Only a few elements are found uncombined in the Earth's crust. Gold is one example. The rest occur as compounds, and have to be extracted from their ores. This is usually carried out by heating with carbon, or by electrolysis.

Some information about the extraction of four different metals is shown below.

Metal	Formula of main ore	Method of extraction
Iron	Fe_2O_3	Heating with carbon
Aluminium	$Al_2O_3 . 2H_2O$	Electrolysis
Copper	Cu_2S	Roasting the ore
Sodium	$NaCl$	Electrolysis

a Give the chemical name of each ore.

b Arrange the four metals in order of reactivity.

c How are the more reactive metals extracted from their ores?

d i How is the least reactive metal extracted from its ore?

ii Why can't this method be used for the more reactive metals?

e Iron and aluminium ores are relatively cheap, but aluminium metal is a lot more expensive than iron metal. Why is this?

f Which of the methods would you use to extract:

i potassium?

ii lead?

iii magnesium?

(Hint: look at the reactivity series on page 148.)

g Gold is a metal found native in the Earth's crust. Explain what *native* means.

h Where should gold go, in your list for b?

i Name another metal which occurs native.

9 Explain why the following metals are suitable for the given uses. (There should be more than one reason in each case.)

a Aluminium for window frames.

b Iron for bridges.

c Copper for electrical wiring.

d Lead for roofing.

e Zinc for coating steel.

10 Many metals are more useful when mixed with other elements than when they are pure.

a What name is given to the mixtures?

b What metals are found in these mixtures? *brass, solder, stainless steel.*

c Describe the useful properties of the mixtures in b.

11 Bauxite is an important raw material. It is the hydrated oxide of a certain metal. The metal is extracted from the oxide by electrolysis.

a Which metal is extracted from bauxite?

b The compound cryolite is also needed for the extraction. Why is this?

c What are the electrodes made of?

d i At which electrode is the metal obtained?

ii Write an equation for the reaction that takes place at this electrode.

e i What product is released at the other electrode?

ii This product reacts with the electrode itself. What problem does that cause?

f Give three uses of the metal obtained from bauxite.

g To improve its resistance to corrosion, the metal is often anodized. How is this carried out? What happens to the surface of the metal?

12 a Draw a diagram of the blast furnace. Try it first without looking back at the diagram on page 158. Show clearly on your diagram:

i where air is 'blasted' into the furnace;

ii where the molten iron is removed;

iii where the second liquid is removed.

b i Name the three raw materials added at the top of the furnace.

ii What is the purpose of each material?

c i What is the name of the second liquid that is removed from the bottom of the furnace?

ii When it solidifies, does it have any uses? If so, name one.

d i Name a waste gas that comes out at the top of the furnace.

ii Does this gas have a use? If so, what?

e Write an equation for the chemical reaction which produces the iron.

f Most of the iron that is obtained from the blast furnace is used to make steel. What element, other than iron, is present in most steel?

13 *aluminium gold iron tin magnesium mild steel calcium stainless steel*

a In the above list of metals and alloys, only four are resistant to corrosion.

i Which are they?

ii Explain why each is resistant to corrosion.

b Which of the other metals or alloys will corrode most quickly? Explain your answer.

11.1 Hydrogen, nitrogen and ammonia

Hydrogen

Hydrogen is the lightest of all elements. On Earth it occurs in many compounds, such as water and methane. But the free element is so light that it has escaped far above the Earth, into the outer atmosphere.

Outside Earth, hydrogen is the most common element in the Universe. The Sun is a white-hot ball of gas, more than a million kilometres across, and largely consists of hydrogen. Its heat and light are produced by nuclear reactions in which hydrogen is converted to helium.

Sunshine – thanks to hydrogen!

The properties of hydrogen

1 It is the lightest of all gases. It is about 20 times lighter than air.
2 It is colourless and has no smell.
3 It is almost insoluble in water.
4 It combines with oxygen to form water:

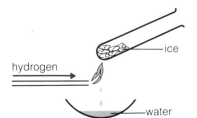

A jet of hydrogen burns in air or oxygen with a blue flame. It can be condensed to water on a cold surface.

A mixture of the two gases explodes when it is lit. The reaction obviously gives out a lot of energy.

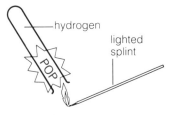

When a lighted splint is held to a test-tube of hydrogen, the gas burns with a squeaky pop. This is used as a test for hydrogen.

In each case, the reaction is:

$$2H_2(g) + O_2(g) \longrightarrow 2H_2O(g)$$

This reaction gives out so much energy that it is used to drive space rockets. The hydrogen and oxygen are stored in tanks in the rockets, in liquid form.

In future, cars may run on hydrogen instead of petrol. Hydrogen forms water when it burns, so would help to cut down pollution.

5 Hydrogen acts as a reducing agent, by removing oxygen. For example, copper(II) oxide is reduced to copper by heating it in a stream of hydrogen. The reaction is:

$$CuO(s) + H_2(g) \longrightarrow Cu(s) + H_2O(g)$$

Notice that the hydrogen is oxidized to water.

Uses of hydrogen

Some is used as fuel.
Some is reacted with vegetable oils, to make margarine.
Some is used to make solvents such as methanol (CH_3OH).

Hydrogen was once used to fill airships but in 1937 an airship called *The Hindenburg* was destroyed, and thirty-five passengers killed, when the hydrogen exploded.

Nitrogen

Nitrogen is all around us – it makes up nearly four-fifths of the air. These are some of its properties:

1 It is a colourless gas, with no smell.
2 It is only slightly soluble in water.
3 It is very unreactive, compared with oxygen.
4 But it combines with hydrogen in the presence of a catalyst to form **ammonia**, NH_3.
5 It also combines with oxygen at high temperatures to form oxides such as **nitrogen monoxide** (NO) and **nitrogen dioxide** (NO_2). This happens in car engines (page 190) and in the air during lightning storms. These oxides are harmful – they can cause us breathing problems and they dissolve in rain to make **acid rain**. But when acid rain soaks into the soil, it can react with substances in the soil to make **nitrates** which plants then use as food.

Nitrogen is essential for all living things, including you. It is needed to make the proteins that are part of every plant and animal. You will learn in biology how it circulates between the air, the soil and living things, in a process called the **nitrogen cycle**.

Ammonia

Ammonia is made by reacting nitrogen with hydrogen, as you will see on the next page. It has the formula NH_3.

1 It is a colourless gas with a strong, choking smell.
2 It is less dense than air.
3 It is easily liquefied by cooling to $-33\,^\circ C$ or by compressing. So it is easy to transport in tanks and cylinders.
4 It reacts with hydrogen chloride gas to form a white smoke. The smoke consists of tiny particles of solid ammonium chloride:

$$NH_3\,(g) + HCl\,(g) \longrightarrow NH_4Cl\,(s)$$

This reaction can be used to test whether a gas is ammonia.
5 It is very soluble in water. (It shows the fountain effect – page 19.)
6 The solution turns red litmus blue – it is **alkaline**. That means it contains hydroxide ions. Some of the ammonia has reacted with water to form ammonium ions and hydroxide ions:

$$NH_3\,(aq) \longrightarrow NH^{4+}\,(aq) + OH^-\,(aq)$$

As only some of its molecules form ions, ammonia is a **weak** alkali.
7 Since ammonia solution is alkaline, it reacts with acids to form salts. For example with nitric acid it forms ammonium nitrate:

$$NH_3\,(aq)^{\cdot} + HNO_3\,(aq) \longrightarrow NH_4NO_3\,(aq)$$

Ammonium nitrate is an important fertilizer.

three shared pairs of electrons

The bonding in nitrogen. Since three pairs of electrons are shared, the bond is a triple bond. It can be shown as N≡N.

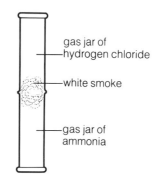

Nitrogen is used as an inert atmosphere in food packages.

gas jar of hydrogen chloride

white smoke

gas jar of ammonia

What is the chemical name for the white smoke?

Questions

1 Why is hydrogen *not* used to fill party balloons?
2 Copy and complete: $FeO\,(s) + H_2\,(g) \longrightarrow \dots$
3 Write down *two* uses of nitrogen in industry
4 Write down three *physical* properties of ammonia.
5 Write an equation to show how ammonia solution reacts with hydrochloric acid.

11.2 Ammonia and nitric acid in industry

Making ammonia in industry

In industry, ammonia is made from nitrogen and hydrogen. The first step is to obtain these gases.

1 **Hydrogen** is made from methane (North Sea gas) and steam, using this reaction:

$$CH_4\,(g) + 2H_2O\,(g) \xrightarrow{\text{catalysts}} CO_2\,(g) + 4H_2\,(g)$$

2 **Nitrogen** is obtained by burning hydrogen in air. Air is mostly nitrogen and oxygen, with small amounts of other gases. Only the oxygen reacts with hydrogen, forming steam:

$$2H_2\,(g) + O_2\,(g) \longrightarrow 2H_2O\,(g)$$

When the steam condenses, the gas that remains is mainly nitrogen.

Part of the ICI ammonia plant at Billingham in Cleveland.

The reaction between them   Nitrogen is unreactive. To make it react with hydrogen, a process called the **Haber process** is used:

1 The two gases are mixed. The mixture is cleaned or **scrubbed**, to get rid of any impurities.
2 Next it is compressed. This pushes the gas molecules closer together.
3 Then it goes to the **converter**. This is a round tank containing beds of hot iron. The iron is a catalyst for this reaction:

$$N_2\,(g) + 3H_2\,(g) \rightleftharpoons 2NH_3\,(g)$$

The reaction is **reversible** (page 122). So it does not go to completion. A mixture of nitrogen, hydrogen and ammonia leaves the converter.
4 This mixture is cooled until the ammonia condenses. Then the nitrogen and hydrogen are pumped back to the catalyst, for another chance to react.
5 The ammonia is run into tanks and stored as a liquid, under pressure.

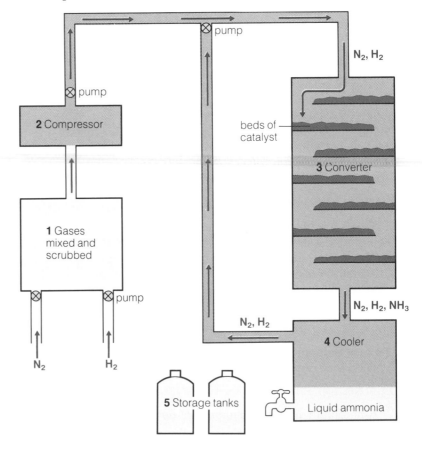

Uses of ammonia

Most of the ammonia from the Haber process is used to make **fertilizers**, such as ammonium nitrate and ammonium sulphate. Some is used to make household cleaners, dyes, explosives and nylon. A lot is used to make nitric acid, as on the next page.

Making nitric acid in industry

A lot of the ammonia from the Haber process is used to make **nitric acid**. The raw materials for nitric acid are **ammonia**, **air** and **water**. This flow chart shows the stages in the process:

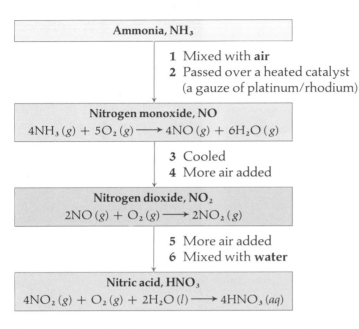

Ammonia, NH₃

    1 Mixed with **air**
    2 Passed over a heated catalyst
       (a gauze of platinum/rhodium)

Nitrogen monoxide, NO
$$4NH_3(g) + 5O_2(g) \longrightarrow 4NO(g) + 6H_2O(g)$$

    3 Cooled
    4 More air added

Nitrogen dioxide, NO₂
$$2NO(g) + O_2(g) \longrightarrow 2NO_2(g)$$

    5 More air added
    6 Mixed with **water**

Nitric acid, HNO₃
$$4NO_2(g) + O_2(g) + 2H_2O(l) \longrightarrow 4HNO_3(aq)$$

Part of the ICI nitric acid plant at Billingham. It is built close to ICI's ammonia plant. Can you explain why?

The overall result is that ammonia is oxidized to nitric acid. Chemical engineers must make sure that no nitrogen monoxide or nitrogen dioxide can escape from the plant, since these gases cause acid rain. Besides, if any nitric acid goes down the drain, it will end up in the river, killing fish and other river life.

Uses of nitric acid

Most of the nitric acid is used to make **fertilizers**.
Some is used to make explosives such as trinitrotoluene (TNT).
Some is used in making nylon and terylene.
Some is used in making drugs.

Questions

1 Ammonia is made from nitrogen and hydrogen.
  a How are the nitrogen and hydrogen obtained?
  b What is the process for making ammonia called?
  c What catalyst is used? What does it do?
  d Write an equation for the reaction.
  e The reaction is *reversible*. What does that mean?
  f What steps are taken to improve the yield of ammonia?

2 Look at the diagram on the opposite page. Why are the catalyst beds arranged that way?
3 a Name the raw materials for making nitric acid.
  b Write word equations for the three reactions that take place during its manufacture.
  c The result of the reactions is that ammonia is *oxidized* to nitric acid. What does that mean?
  d What is the main use of nitric acid?

11.3 Fertilizers

What plants need

In order to grow, a plant needs carbon dioxide, light and water. It also needs several different elements. The main ones are **nitrogen**, **potassium** and **phosphorus**.

Nitrogen is essential for proteins, which make strong stems and healthy leaves.

Potassium helps a plant to survive frost and to resist disease.

Phosphorus is needed to help roots develop. It also helps crops ripen.

Plants also need smaller amounts of **calcium**, **magnesium**, **sodium** and **sulphur**, and tiny amounts of **copper**, **iron**, **zinc**, **manganese** and **boron**. They obtain all these elements from compounds in the soil, which they take in through their roots as solutions.
But what happens when a farmer grows the same crops over and over again in the same field? The compounds in the soil all get used up, and the crops start to suffer. That's where fertilizers come in.

Fertilizers

Fertilizers are substances added to soil to replace the elements used up by plants. Animal manure is a **natural fertilizer**. But most farmers use at least some **artificial fertilizers**. These are chemicals such as:

ammonium nitrate, NH_4NO_3	ammonium sulphate, $(NH_4)_2SO_4$
calcium nitrate, $Ca(NO_3)_2$	potassium sulphate, K_2SO_4
ammonia solution, NH_3 (*aq*)	ammonium phosphate, $(NH_4)_3PO_4$

You can see that ammonium phosphate will put nitrogen and phosphorus back into the soil. What about potassium sulphate?

Choosing a fertilizer   The fertilizers on sale are usually a mixture of chemicals. They are made up to suit different soils and crops. For example, some contain more nitrogen than others. Farmers can have their soil tested to find out what it's short of. Then they can buy a fertilizer to suit.
Suppose a farmer wants to grow grass to feed cattle. Grass needs a lot of nitrogen. So if the soil is low in nitrogen, a fertilizer mixture high in nitrogen would be the best one to buy.

The fertilizer mixture in this bag contains 20% nitrogen, 8% phosphorus and 14% potassium.

A machine spreading a slurry of animal manure. It contains the natural fertilizer urea, amongst others.

Cattle farmers use nitrogen fertilizers to help their grass grow.

The problem with fertilizers

Fertilizers allow farmers to grow more crops in the same soil year after year. That sounds good, but there are problems.

First, crops grown using artificial fertilizer are often tasteless and low in nutrients. They don't stay fresh long. And they may be contaminated by the fertilizer. For example, nitrates from fertilizer can collect in lettuce. Nitrates have been linked with disease in babies and with stomach cancer.

Second, fertilizers get washed out of the soil by rain and end up in river water. There they fertilize tiny green water plants called **algae**. These grow so well that they cover the water. So plants below the surface die because no light can reach them. Bacteria feed on the dead plants, using up dissolved oxygen as they do so. And soon there isn't enough oxygen left in the river for fish and other living things. They die from oxygen starvation.

An example of 'choking' by the uncontrolled growth of algae, caused by fertilizer.

Third, nitrates that get into the river can then pass through the waterworks and end up in our drinking water. As you saw above, they have been linked with health problems.

And finally, heavy use of fertilizers upsets the balance of nature. Fertilizers encourage weeds as well as crops. The crops and weeds attract more insects. So the farmer ends up using lots of weedkiller and pesticide. These kill birds and other wildlife, and they are also bad for humans.

Questions

1 Name the three main elements that plants need, and say why they need them.
2 What are *fertilizers* and why are they used?
3 Write down as many points as you can to show:
   a the benefits of using animal manure as fertilizer;
   b the problems that could arise.
4 What does *organic farming* mean?

5 What problems might there be in using ammonia solution as a fertilizer?
6 Imagine you are a farmer. List all the good points you can about using artificial fertilizers.
7 Now list the problems caused by fertilizers. Put them into what *you* feel is their order of importance, with the worst problem first.

11.4 A look at a fertilizer factory

What the factory makes

A modern fertilizer factory will make two main kinds of fertilizer:

1 **Straight N fertilizer** for farmers who want only nitrogen. This is usually in the form of ammonium nitrate, NH_4NO_3.
2 **NPK compound fertilizers** for farmers who want nitrogen, phosphorus, and potassium. These are usually a mixture of ammonium nitrate, ammonium phosphate, and potassium chloride.

The factory may make other fertilizers too, but in very much smaller amounts.

The production process

This diagram shows the raw materials the factory needs to make 1000 tonnes each of straight and compound fertilizer a day, and the steps in the production process. (The term 't/day' stands for 'tonnes per day'.)

Fertilizers are made all year round, but sales are seasonal. A fertilizer pellet could lie buried in a 20-metre heap for several months. It must be able to survive storage without losing shape or sticking to other pellets.

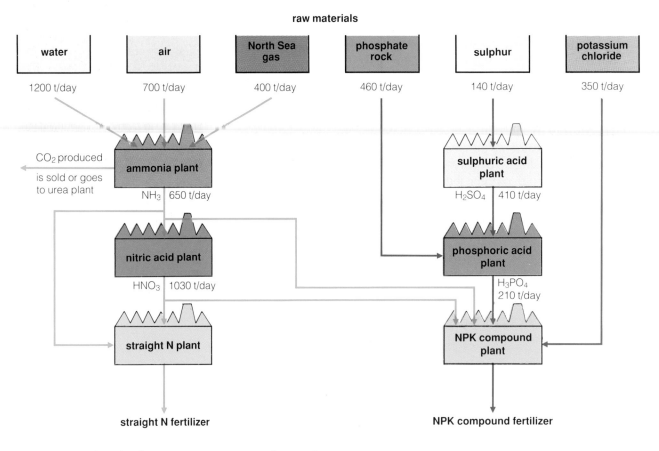

raw materials

| water | air | North Sea gas | phosphate rock | sulphur | potassium chloride |
| 1200 t/day | 700 t/day | 400 t/day | 460 t/day | 140 t/day | 350 t/day |

CO_2 produced is sold or goes to urea plant

ammonia plant

NH_3 650 t/day

nitric acid plant

HNO_3 1030 t/day

straight N plant

straight N fertilizer

sulphuric acid plant

H_2SO_4 410 t/day

phosphoric acid plant

H_3PO_4 210 t/day

NPK compound plant

NPK compound fertilizer

You can see that the factory is not just a single unit. It is **six separate plants** built close together on the same site. Each plant is controlled so that it is making the right amount of a substance at the right time.

Mollie Travis, personnel officer: 'Of course any new factory has to staff up well in advance of production. We were running training courses for local workers six months before we opened.'

Michael Nelson, chemical engineer at the factory: 'My job is mostly to do with the environment. We continually monitor the waste gases and liquids leaving the plant to make sure they're safe.'

Graham Taylor, technician: 'Me? I've just come off the night shift. A fertilizer factory like this one keeps going 24 hours a day, every day. It's got to, to be economical.'

Choosing the factory site

A company considering a new site for a fertilizer factory needs to think about things like these:

1 **Closeness to raw materials.** For example, potassium chloride is mined near Whitby in Yorkshire. Should we build near there?
2 **Closeness to ports.** We need to import sulphur and phosphate. We want to export fertilizer. Should we be close to a port?
3 **The road network.** We will be delivering fertilizer all over the country. Are we close to major roads?
4 **The water supply.** We need hundreds of tonnes of water a day as a raw material, and for cooling. Is it available here?
5 **The gas supply.** We need North Sea gas as a raw material, and as a fuel. Any problems getting it here?
6 **The workforce.** Can we get the kinds of worker we need locally? Will we be able to attract key staff from other areas?
7 **The cost of land.** Is there somewhere cheaper we should go?
8 **Government grants.** Are there special grants for starting a factory in this area?
9 **Objections from the local people.** They may worry that the factory will pollute the air, or the river, or make too much noise, or spoil a view. Will we be able to set their minds at rest?

The company is not likely to find a perfect site. It will have to balance all these factors and choose the site that fits best.

Emily Benson, chemist: 'In fact there are strict legal requirements for the amount of N, P and K in fertilizers. We have machines that check the contents automatically. But we do lab checks too.'

Questions

You may need to look back at pages 166 and 167 to answer some of these questions.

1 What are the two main types of fertilizer a factory is likely to make? Which chemicals do these usually contain?
2 List the raw materials the factory needs.
3 List the acids the factory makes.
4 What is the sulphur used for?

5 North Sea gas is used as a raw material. For which process? Write an equation for the reaction.
6 Write word equations for the reactions involved in making nitric acid from ammonia.
7 List all the ways a fertilizer factory might cause pollution.
8 When choosing a factory site, which *four* factors do you think are most important? Explain your choice.

11.5 Oxygen

Making oxygen in the laboratory

The diagram on the right shows one way to make oxygen in the laboratory.

Hydrogen peroxide is a clear colourless liquid, with the formula H_2O_2. It decomposes to water and oxygen, like this:

$$2H_2O_2\,(aq) \longrightarrow 2H_2O\,(l) + O_2\,(g)$$

The reaction is very slow, so black manganese(IV) oxide is added as a catalyst.

Oxygen is not very soluble in water. How can you tell, from the apparatus?

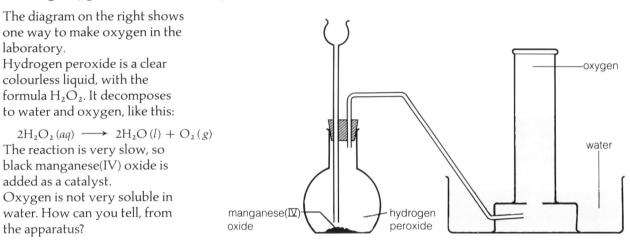

oxygen

water

manganese(IV) oxide

hydrogen peroxide

The properties of oxygen

1 It is a clear, colourless gas, with no smell.
2 It is only slightly soluble in water.
3 It is very reactive. It reacts with a great many substances to produce **oxides**, and the reactions usually give out a lot of energy. For us, its two most important reactions are **respiration** and the **combustion of fuels**. These are very similar, as you will see.

Respiration   This is the process that keeps us alive. During respiration, oxygen reacts with glucose in our bodies. The reaction produces carbon dioxide, water, and the energy we need:

glucose   + oxygen ⟶ carbon dioxide +   water   + energy
$C_6H_{12}O_6\,(aq)$ + $6O_2\,(g)$ ⟶ $6CO_2\,(g)$ + $6H_2O\,(g)$ + energy

The glucose comes from digested food, and the oxygen from air:

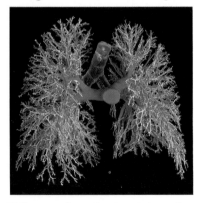

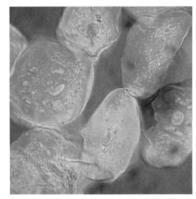

When we breathe in, air travels to our lungs. It passes along tiny tubes to the lungs' surface. The oxygen diffuses through the surface, into the blood.

The blood carries it to the million of cells in the body, along with glucose from digested food. Respiration takes place in each cell.

The energy from respiration keeps our heart and muscles working. It also keeps us warm. Without it, no body reactions could go on. We would die.

The carbon dioxide and water from respiration pass from the cells back into the blood. The blood carries them to the lungs, and we breathe them out.

Respiration goes on in the cells of *all* living things. Fish use the oxygen dissolved in water, which they take in through their gills. Plants use oxygen from the air, and take it in through tiny holes in their leaves.

Combustion of fuels   Fuels are substances we burn to get energy – usually in the form of heat. The burning needs oxygen:

North Sea gas is a fuel. It is mainly **methane**. It is pumped into homes from gas wells in the North Sea.

In the pipes of gas cookers, fires and boilers, the methane is mixed with air. When the mixture is lit, the methane . . .

. . . reacts with the oxygen in the air, giving out energy as heat and light. The heat is used to cook, heat homes and water.

The equation for the reaction is:

$$\text{methane} + \text{oxygen} \longrightarrow \text{carbon dioxide} + \text{water} + \text{energy}$$
$$CH_4\,(g) + 2O_2\,(g) \longrightarrow CO_2\,(g) + 2H_2O\,(g) + \text{energy}$$

Compare this with the equation for respiration. What do you notice?

Petrol, oil, coal and wood are also used as fuels. Each is a mixture of compounds containing carbon and hydrogen. On burning in plenty of oxygen, they all produce carbon dioxide, water and energy.

If there is only a limited amount of oxygen, **carbon monoxide** (CO) is produced instead of carbon dioxide, when fuels burn. This is a problem, because carbon monoxide is a deadly poisonous gas.

Test for oxygen   Things burn much faster in pure oxygen than in air. The reason is that the oxygen in air is **diluted** by nitrogen and other gases. This gives us a way to test a gas, to see if it is oxygen:

1 A wooden splint is lit. Then the flame is blown out. The splint keeps on glowing, because the wood is reacting with oxygen.
2 The glowing splint is plunged into the unknown gas.
3 If the gas is oxygen, the splint immediately bursts into flame.

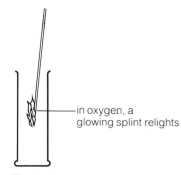

in oxygen, a glowing splint relights

The test for oxygen.

Questions

1 Draw a labelled diagram, showing how you would make oxygen from hydrogen peroxide.
2 This method produces 'damp' oxygen. Why? Can you suggest a way to dry it?
3 What is respiration? Where does it take place?

4 How do fish obtain oxygen for respiration?
5 What is a fuel? Give four examples.
6 Write down the equations for respiration and the burning of methane. In what ways are they alike?
7 How would you test a gas to see if it was oxygen?

11.6 Oxides

On page 172 you saw that oxygen reacts with many substances to form **oxides**. There are different types of oxide, as you will see below.

Basic oxides

Look at the way these metals react with oxygen:

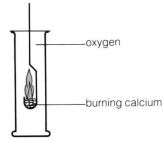

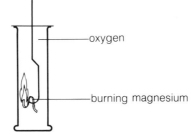

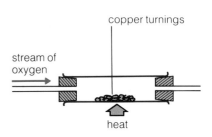

Some grains of calcium are lit over a Bunsen flame, then plunged into a jar of oxygen. They burn brightly, leaving a white solid called **calcium oxide**:

$$2Ca\,(s) + O_2\,(g) \longrightarrow 2CaO\,(s)$$

Magnesium ribbon is lit over a Bunsen flame, and plunged into a jar of oxygen. It burns with a brilliant white flame, leaving a white ash called **magnesium oxide**:

$$2Mg\,(s) + O_2\,(g) \longrightarrow 2MgO\,(s)$$

Copper is too unreactive to catch fire in oxygen. But when it is heated in a stream of the gas, its surface turns black. The black substance is **copper(II) oxide**:

$$2Cu\,(s) + O_2\,(g) \longrightarrow 2CuO\,(s)$$

The way each metal reacts depends on its **reactivity**. The last reaction above produces copper(II) oxide, which is insoluble in water. But it does dissolve in dilute acid:

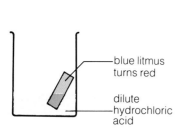

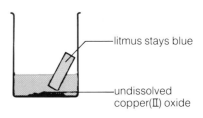

This is dilute hydrochloric acid. It turns blue litmus paper red, as all acids do.

Copper(II) oxide dissolves in it, when it is warmed. But after a time, no more will dissolve.

The resulting liquid has no effect on blue litmus. So the oxide has **neutralized** the acid.

Copper(II) oxide is called a **base**, or **basic oxide**, since it can neutralize acid. The products are a salt and water:

$$\text{base} + \text{acid} \longrightarrow \text{salt} + \text{water}$$
$$CuO\,(s) + 2HCl\,(aq) \longrightarrow CuCl_2\,(aq) + H_2O\,(l)$$

Calcium oxide and magnesium oxide behave in the same way – they too can neutralize acid, so they are basic oxides.
In general, metals react with oxygen to form basic oxides.

Acidic oxides

Now look at the way these nonmetals react with oxygen:

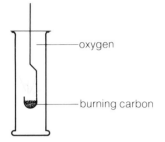

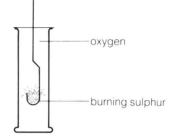

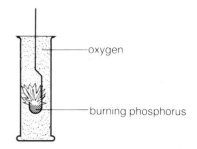

Carbon powder is heated over a Bunsen, until it is red-hot. It is then plunged into a jar of oxygen. It glows bright red and the gas **carbon dioxide** is formed:

$$C\,(s) + O_2\,(g) \longrightarrow CO_2\,(g)$$

Sulphur catches fire over a Bunsen, and burns with a blue flame. In pure oxygen it burns even more brightly. The product is **sulphur dioxide**, a gas:

$$S\,(s) + O_2\,(g) \longrightarrow SO_2\,(g)$$

Phosphorus bursts into flame in air or oxygen, without being heated. (That is why it is stored under water.) A white solid called **phosphorus pentoxide** is formed:

$$P_4\,(s) + 5O_2\,(g) \longrightarrow P_4O_{10}\,(s)$$

The first reaction above produces carbon dioxide, which is slightly soluble in water:

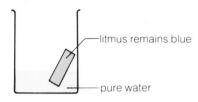

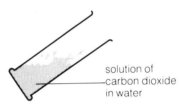

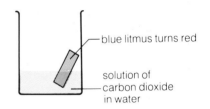

This is pure water. It is neutral. That means it has no effect on red *or* blue litmus paper.

When it is poured into the jar of carbon dioxide, and shaken well, it dissolves some of the gas.

The solution turns blue litmus paper red, so it is **acidic**. It is called **carbonic acid**.

Carbon dioxide is called an **acidic oxide**, because it dissolves to form an acid. Sulphur dioxide and phosphorus pentoxide also dissolve in water to form acids, so they too are acidic oxides.
In general, nonmetals react with oxygen to form acidic oxides.

Neutral oxides

Nitrogen forms the acidic oxide **nitrogen dioxide**, NO_2. But it also forms an oxide which is not acidic, called **dinitrogen oxide**, N_2O. Dinitrogen oxide is quite soluble in water, but the solution is neutral — it has no effect on litmus. So dinitrogen oxide is a **neutral oxide**. Another example of a neutral oxide is carbon monoxide, CO.

These oxides are unusual because they react with acids *and* bases:
Aluminium oxide Al_2O_3
Zinc oxide   ZnO
Lead(II) oxide   PbO
They are called **amphoteric** oxides.

Questions

1 How could you show that magnesium oxide is a base?
2 Copy and complete: Metals usually form .... oxides while nonmetals form .... oxides.
3 Why is phosphorus stored under water?
4 What colour change would you see, on adding litmus solution to a solution of phosphorus pentoxide?
5 Name two nonmetal oxides that are *neutral*.

11.7 Sulphur and sulphur dioxide

Sulphur

Sulphur is obtained from two sources:

1 Most comes from large underground sulphur beds found in Poland, Mexico, and the USA.
2 Some is extracted from crude oil and natural gas.

Some crude oil and natural gas contains a high percentage of sulphur. If the oil or gas is burned, sulphur dioxide forms. This causes air pollution and acid rain. So oil companies are forced to remove the sulphur before the oil or gas can be used.

The properties of sulphur

1 It is a brittle yellow solid.
2 It is made up of crown-shaped molecules, each with eight atoms.
3 Because it is molecular, it has quite a low melting point (115 °C). It melts easily in a Bunsen flame.
4 Like other nonmetals, it does not conduct electricity.
5 Like most nonmetals, it is insoluble in water.
6 It reacts with metals to form sulphides. With iron it forms iron(II) sulphide:

$$Fe\ (s) + S\ (s) \longrightarrow FeS\ (s)$$

7 It burns in oxygen to form sulphur dioxide:

$$S\ (s) + O_2\ (g) \longrightarrow SO_2\ (g)$$

A molecule of sulphur.

Uses of sulphur

Most of it is used to make sulphuric acid.
Some is added to rubber, to make it tough and strong.
Some is used to make drugs, pesticides, matches and paper.
Some is used to make a special concrete called **sulphur concrete**.
Unlike ordinary concrete, sulphur concrete is not attacked by acid.
So it is used for floors and walls in factories where acid may get spilled.

Sulphur has two crystalline forms or allotropes. This is a crystal of **rhombic** sulphur, the form which is stable at room temperature.

But if you heat rhombic sulphur to above 96 °C, the molecules rearrange themselves to form needle-shaped crystals of **monoclinic** sulphur.

Sulphur concrete used for the floor of a copper electroplating plant. Sulphuric acid used for the electroplating process would corrode ordinary concrete.

Sulphur dioxide

Sulphur dioxide is formed when sulphur burns in air. Its formula is SO_2. It has these properties:

1 It is a colourless gas, with a strong, choking smell.
2 It is heavier than air.
3 It is soluble in water. The solution is acidic because the gas reacts with water to form **sulphurous acid**, H_2SO_3:

$$H_2O\,(l) + SO_2\,(g) \longrightarrow H_2SO_3\,(aq)$$

Sulphur dioxde is therefore an acidic oxide. The acid easily decomposes again to sulphur dioxide and water.
4 It acts as a bleach when it is damp or in solution. Some coloured things lose colour when they lose oxygen — that is, when they are **reduced**. Sulphur dioxide bleaches them by reducing them.
5 When it escapes into the air from engine exhausts and factory chimneys, it causes air pollution. It attacks the breathing system in humans and other animals. It dissolves in rain to give acid rain. Acid rain damages buildings, metalwork, and plants.

These trees were killed by sulphur dioxide, which dissolved in rain to give *acid rain*.

Uses of sulphur dioxide

Some is used to bleach wool, silk, and wood pulp for making newspaper. Some is used in the preparation of soft drinks, jam, and dried fruit, because it stops the growth of bacteria and moulds. But most is used to make sulphuric acid, as shown on page 178.

Scrap paper being recycled for newsprint. The next stage is to bleach the paper using sulphur dioxide.

This photograph shows two sets of potato strips. The strips on the left have been treated with sulphur dioxide to stop browning.

Questions

1 What are the two sources of sulphur?
2 Sulphur reacts with lead to form lead(II) sulphide. Write an equation for the reaction.
3 Write down three uses of sulphur.
4 Explain why sulphur dioxide makes rain acidic.
5 Sulphur dioxide is heavier than air. How do you think that might affect pollution?
6 Write down three uses of sulphur dioxide.

177

11.8 Sulphuric acid and sulphates

How sulphuric acid is made

More sulphuric acid is produced each year than any other chemical. Most of it is made by the Contact process. The raw materials are sulphur, air and water. This flow chart shows what happens:

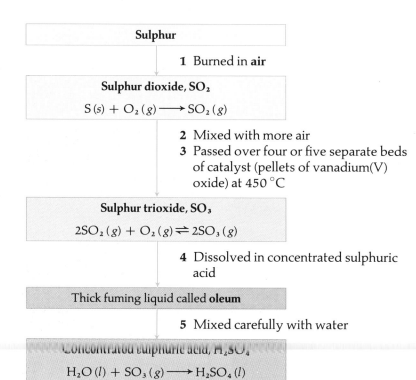

Sulphur

1 Burned in **air**

Sulphur dioxide, SO₂
$S(s) + O_2(g) \longrightarrow SO_2(g)$

2 Mixed with more air
3 Passed over four or five separate beds of catalyst (pellets of vanadium(V) oxide) at 450 °C

Sulphur trioxide, SO₃
$2SO_2(g) + O_2(g) \rightleftharpoons 2SO_3(g)$

4 Dissolved in concentrated sulphuric acid

Thick fuming liquid called **oleum**

5 Mixed carefully with water

Concentrated sulphuric acid, H₂SO₄
$H_2O(l) + SO_3(g) \longrightarrow H_2SO_4(l)$

Part of the ICI sulphuric acid plant at Billingham.

This warning sign is displayed on all containers and tankers containing sulphuric acid. What message does it give?

Things to note about the Contact process

1 The burning of sulphur in air is exothermic (it gives out heat). So is the reaction between sulphur dioxide and oxygen.

2 The reaction between sulphur dioxide and oxygen is reversible (see page 122). That means some sulphur trioxide is continually breaking down again to sulphur dioxide and oxygen. So the mixture is passed over several beds of catalyst to give the two gases another chance to react.

3 The catalyst will not work at all below 400 °C. It works better as the temperature rises. But the yield of sulphur trioxide falls as the temperature rises, since the reaction is exothermic. So 450 °C is chosen as a compromise, to give a good yield quickly.

4 The heat given out by the reactions is used to make steam. The steam is used to make electricity, or sold to nearby factories for heating. This helps to cover the cost of running the plant.

Sulphuric acid is a hazardous chemical to store and transport. If possible, the acid plant is built close to major users, to keep transport costs down.

Uses of sulphuric acid

Most of it is used to make fertilizers such as ammonium sulphate. Some is used to make paint, soapless detergents, soap, dyes, and plastic. Some is used as battery acid. In fact nearly every industry uses some sulphuric acid. (It is the cheapest acid to buy.)

The properties of sulphuric acid

The concentrated acid   It is a colourless oily liquid. It is a **dehydrating agent**: it can remove water. It will dehydrate sugar, paper and wood. These are all made of carbon, hydrogen and oxygen. The acid removes the hydrogen and oxygen as water, leaving carbon behind. For example:

$$C_6H_{12}O_6 \xrightarrow[H_2SO_4]{conc} 6C + 6H_2O$$

$$\text{sugar} \longrightarrow \text{carbon} + \text{water}$$

It will dehydrate flesh in the same way, so it's very dangerous!

The dilute acid   It has typical acid properties:

1 It turns blue litmus red.
2 It reacts with metals to give hydrogen, and salts called **sulphates**.
3 It reacts with metal oxides and hydroxides to give sulphates and water.
4 It reacts with carbonates to give sulphates, water, and carbon dioxide.

The dilute acid is made by carefully adding concentrated acid to water (never the other way round, because so much heat is produced that the acid could splash out and burn you).

Test for sulphates

Most sulphates are soluble in water. But lead, calcium and barium sulphates are insoluble. So to see if a solution contains a sulphate:

1 First add a few drops of dilute hydrochloric acid, to acidify it.
2 Then add a few drops of barium chloride solution.
3 If a sulphate is present, a white precipitate forms. The white precipitate is barium sulphate, $BaSO_4$.

This is what happened when concentrated sulphuric acid was added to two teaspoons of sugar. It dehydrated the sugar, leaving carbon behind.

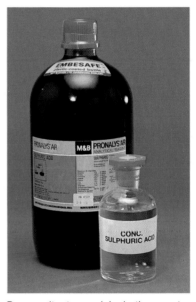

Because it acts as a dehydrating agent, concentrated sulphuric acid can also be used in the lab to dry gases, as long as it does not react with them.

Questions

The information on page 122 about reversible reactions will help you answer these questions.
1 For making sulphuric acid, name: **a** the process; **b** the raw materials; **c** the catalyst.
2 The reaction between sulphur dioxide and oxygen is *reversible*. What does that mean?
3 **a** Is the *breakdown* of sulphur trioxide to sulphur dioxide endothermic or exothermic?
   b Does *more* sulphur trioxide break down, or less, as the temperature rises? Why?

4 **a** Why is a catalyst needed?
   b At 500 °C, the catalyst makes sulphur trioxide form even faster. Why isn't this temperature used?
5 Explain how this helps to increase the yield of sulphur trioxide:
   a Several beds of catalyst are used.
   b The sulphur trioxide is removed by dissolving it.
6 Copy and complete for the test for sulphates:
   Ba^+ (*aq*) + ... $\longrightarrow$ ....

11.9 Chlorine and its compounds

Chlorine

Chlorine is not found as the free element in nature. It occurs mainly as the compound sodium chloride or **rock salt**. It is obtained by electrolysing either molten salt or **brine**, a concentrated solution of salt in water. (See pages 102 and 106.) It has these properties:

1 It is a greenish-yellow gas with a choking smell.
2 It is heavier than air.
3 It is poisonous to all living things.
4 It is soluble in water. The solution is called **chlorine water**. It is acidic because chlorine reacts with water to form *two* acids:

$$Cl_2\,(g) + H_2O\,(l) \longrightarrow \underset{\text{hydrochloric acid}}{HCl\,(aq)} + \underset{\text{hypochlorous acid}}{HOCl\,(aq)}$$

Hypochlorous acid slowly decomposes again, giving off oxygen:

$$2HOCl\,(aq) \longrightarrow 2HCl\,(aq) + O_2\,(g)$$

5 Chlorine water acts as a bleach. This is because the hypochlorous acid in it can lose oxygen to other substances – it can **oxidize** them. Some coloured substances turn colourless when they are oxidized.
6 Like other bleaches, chlorine water also acts as a sterilizing agent – it kills bacteria and other germs.
7 Chlorine reacts with many nonmetals and most metals. For example, hydrogen burns in chlorine to form hydrogen chloride. The reaction can be explosive.

$$H_2\,(g) + Cl_2\,(g) \longrightarrow 2HCl\,(g)$$

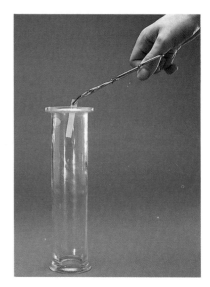

Testing for chlorine in the laboratory. The gas bleaches damp litmus. Can you explain why?

When chlorine is passed over heated aluminium, the metal glows white and turns into aluminium(III) chloride:

$$2Al\,(s) + 3Cl_2\,(g) \longrightarrow 2AlCl_3\,(s)$$

8 Chlorine is more reactive than the halogens below it in Group 7. So it will **displace** them from solutions of their compounds. Potassium bromide solution is colourless, but turns orange when you bubble chlorine through it because the bromine is displaced:

$$\underset{\text{colourless}}{2KBr\,(aq)} + Cl_2\,(g) \longrightarrow 2KCl\,(aq) + \underset{\text{orange}}{Br_2\,(aq)}$$

A solution of potassium iodide also changes colour:

$$\underset{\text{colourless}}{2KI\,(aq)} + Cl_2\,(g) \longrightarrow 2KCl\,(aq) + \underset{\text{red-brown}}{I_2\,(aq)}$$

Uses of chlorine

It is used to sterilize drinking water and the water in swimming pools, and to make polyvinylchloride (PVC), bleaches, pesticides, weedkillers, and solvents such as tetrachloroethane (used in dry-cleaning). Some is turned into hydrochloric acid.

Because chlorine is poisonous, it was used as a weapon in World War I. This soldier is ready for a chlorine gas attack.

Hydrogen chloride

Hydrogen chloride is made in industry by burning hydrogen in chlorine. The gas has these properties:

1 It is heavier than air.
2 It has a choking smell, and it irritates the eyes and lungs.
3 It dissolves very easily in water, to form **hydrochloric acid**. Like all acids, this contains hydrogen ions:

$$HCl\,(aq) \longrightarrow H^+\,(aq) + Cl^-\,(aq)$$

4 Hydrogen chloride reacts with ammonia to form white smoke, made from tiny particles of solid ammonium chloride (page 165):

$$NH_3\,(g) + HCl\,(g) \longrightarrow NH_4Cl\,(s)$$

This reaction is used to test for ammonia, or hydrogen chloride.

You can test for ammonia, using a glass rod dipped in concentrated hydrochloric acid. The white fumes prove that the gas coming from the beaker is ammonia.

Hydrochloric acid

Hydrochloric acid is made in industry by dissolving hydrogen chloride in water. It is a typical acid. It reacts:

● with metals to give hydrogen and salts called **chlorides**.
● with metal oxides and hydroxides to form chlorides and water.
● with carbonates to form chlorides, water and carbon dioxide.

Sodium chloride

Sodium chloride (NaCl) is the most important chloride of all.

1 It occurs naturally as rock salt, and in sea water.
2 It is the starting point for many other chemicals. For example, electrolysis of brine gives sodium hydroxide, chlorine and hydrogen and is the basis of the chlor-alkali industry (page 106).
3 It improves the flavour of food. Sodium ions are essential for body fluids. But too much salt can cause high blood-pressure.
4 It is used to melt ice on the roads in winter. But it also makes objects made from iron and steel rust more quickly.

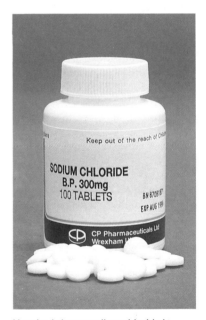

Your body loses sodium chloride in sweat. The result can be exhaustion. So people in hot climates often have to take salt tablets.

Test for chlorides

Only silver and lead chlorides are insoluble in water. So, to see if an unknown solution contains a chloride, this is what to do:

1 First add a few drops of nitric acid, to acidify the solution.
2 Then add a little silver nitrate solution.
3 A chloride will make a white precipitate of silver chloride:

$$Ag^+\,(aq) + Cl^-\,(aq) \longrightarrow AgCl\,(s).$$

Questions

1 Explain why a solution of chlorine in water:
  a is acidic; b is able to bleach things.
2 Write equations to show how chlorine reacts with:
  a sodium; b iron; c sodium iodide solution.
3 Write equations to show how hydrochloric acid reacts with: a sodium hydroxide; b zinc.
4 Describe how you would test a solution to see if it contains a chloride.

Questions on Section 11

1 The manufacture of ammonia and nitric acid are both very important industrial processes.

A Ammonia
a Name the raw materials used.
b Which two gases react together?
c Why are the two gases scrubbed?
d Why is the mixture passed over iron?
e What happens to the *unreacted* nitrogen and hydrogen?
f Why is the manufactured ammonia stored at high pressure?

B Nitric acid
a Name the raw materials used.
b Which chemicals react together to form nitric acid?
c What would happen if the gauze containing platinum and rhodium was removed?
d Why must the chemical plant be constantly checked for leaks?

2 Write equations for the chemical reactions in question 1.

3 Using the apparatus below, dry ammonia is passed over heated copper(II) oxide. The gases given off are passed through a cooled U-tube. A liquid (A) forms in the U-tube and a colourless gas (B) collects in the gas jar.
a The copper(II) oxide is reduced to copper. What would you see as the gas passes over the heated copper oxide?
b Why is the U-tube surrounded by a freezing mixture?
c The liquid A is found to have a boiling point of 100 °C. Identify liquid A.
d Is gas B soluble or insoluble in water?
e Identify gas B. (Hint: look at the other chemicals.)
f Write a word equation for the reaction.
g The copper oxide is *reduced* by the ammonia. Explain what this means.
h How will the mass of the heated tube and contents change during the reaction?

4 The diagram shows apparatus for an experiment to make ammonia. X and Y are two gas syringes, connected to a combustion tube A. At the start, syringe X contained 75 cm³ of hydrogen and syringe Y contained 25 cm³ of nitrogen.

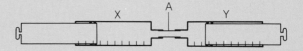

a Copy out the diagram, labelling the gases.
b How would you make the gases mix?
c The reaction needs a catalyst. Why?
d Name a suitable catalyst.
e Where should the catalyst be placed? Add the catalyst to your diagram.
f Where should the apparatus be heated? Show this on your diagram.
g How would you show that some ammonia had been obtained, at the end of the experiment?

5 Ammonium compounds and nitrates are of great importance as fertilizers.
a Why do these compounds help plant growth?
b Name one *natural* fertilizer.
c Name two compounds containing nitrogen which are manufactured for use as fertilizers.
Write the chemical formulae for these compounds.
d Name two elements other than nitrogen which plants need, and explain their importance to the plants.
e Why are some fertilizers not suitable for quick-growing vegetables like lettuce?
f Some fertilizers are acidic. What is usually added to soils to correct the level of acidity?
g Land which is intensively farmed needs regular applications of fertilizer. Explain why.
h Fertilizers obviously have advantages. But many people are worried about the increasing use of fertilizers, especially nitrates, by farmers. Can you suggest why?

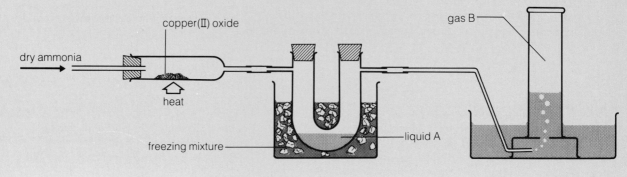

6 Look at each description below in turn. Say whether it fits oxygen, or sulphur, or chlorine.
  a Quite soluble in water.
  b Solid at room temperature.
  c Reacts with metals to form oxides.
  d Exists in more than one solid form.
  e When damp, removes the colour from dyes.
  f Burns in air with a blue flame.
  g Reacts with hydrogen to form water.
  h A poisonous gas.
  i Is added to rubber to make it tough and strong.
  j Relights a glowing splint.
  k Is colourless.
  l Reacts with other elements to form chlorides.
  m Forms a gaseous oxide which causes acid rain when burnt.

7 The elements sulphur and oxygen are in the same group of the periodic table, so they have similar properties. But there are also some differences between them. Use the information on pages 172–177 to answer these questions.
  a Oxygen is a nonmetal. Is sulphur a nonmetal?
  b Do the elements look alike at room temperature? Explain your answer.
  c Both the elements are molecular. What type of bonding do their molecules contain?
  d In what way are their molecules different?
  e Sulphur combines with hydrogen to form the gas hydrogen sulphide, H_2S. Name one way in which this differs from the compound formed between oxygen and hydrogen.

8 The following diagrams represent molecules of different substances that contain sulphur.
  a Write the chemical formula for each substance, then name the substance.

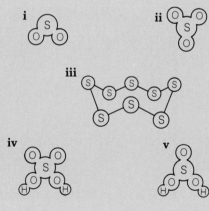

  b How would you convert:
    substance **iii** into substance **i**?
    substance **i** into substance **ii**?
    substance **i** into substance **v**?

9 This table shows the properties of certain oxides.

| Oxide | State at 20 °C | When added to water: | |
		Is energy given out?	pH of solution
Magnesium oxide	solid	no	above 7
Calcium oxide	solid	yes	above 7
Copper(II) oxide	solid	no	–
Sulphur dioxide	gas	yes	below 7
Carbon dioxide	gas	no	below 7

  a Why is no pH given for copper(II) oxide?
  b Which two compounds react when added to water? Give a reason for your choice.
  c Which two compounds contain only nonmetals?
  d What conclusions can you draw about:
    i the pH of solutions of oxides?
    ii the state of oxides at room temperature?

10 Below is a flow chart for the Contact process.

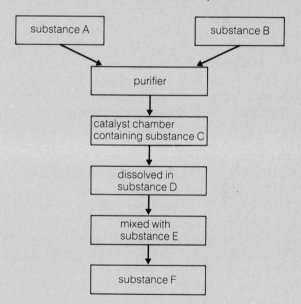

  a Name substances A, B, C, D, E and F.
  b Why is a catalyst used?
  c Write a chemical equation for the reaction that takes place on the catalyst.
  d Why is the production of substance F very important? Give three reasons.
  e Copy out the flow chart, writing in the full names of the different substances.

11 When a piece of damp blue litmus paper is placed in a gas jar of chlorine, the litmus changes colour. First the blue colour turns to red, then the red colour quickly disappears, leaving the paper white.
  a Explain why the paper turns red.
  b Explain why the red colour then disappears.

12.1 Carbon and some carbon compounds

Carbon

A small amount of carbon occurs as the free element in the Earth's crust. It occurs in two forms, **diamond** and **graphite**. These are the **allotropes** of carbon (page 57).
Diamond is a very hard, clear solid that can be cut and polished so that it sparkles in light. Graphite is a dark, greasy solid.
Charcoal and soot are forms of graphite. They are made by heating coal, wood or animal bones in just a little air.

Carbon compounds   Carbon also occurs naturally in thousands of compounds.

1 It is found as **carbon dioxide** in the air.
2 It is contained in **proteins, carbohydrates** and other compounds that make up living things.
3 Coal, wood and natural gas are mixtures of carbon compounds.
4 It occurs in many rocks, as **carbonates**.

Diamond is so hard that it can be used to cut stone. This cutting wheel is edged with diamond.

Carbonates

Carbonates are compounds that contain the carbonate ion, CO_3^{2-}.
Sodium carbonate has the formula Na_2CO_3. It is often called **washing soda**, and it is used in making glass, and to soften water.
Calcium carbonate has the formula $CaCO_3$, and occurs naturally as **limestone, chalk** and **marble**. Limestone is the most common of these.

Uses of limestone   There are many.

1 It is cut into blocks and used for building.
2 It is used for extracting **iron**, in the blast furnace (page 158).
3 It is heated with sand and sodium carbonate, to make **glass**.
4 It is heated with clay to make a grey powder called **cement**. When water is added, cement sets to a hard mass.
5 It is used for the tiny stone chips (aggregate) in concrete.
6 It is a base, which means it can be used to neutralize acids, and it is cheap. So farmers sometimes spread powdered limestone on soil that is too acidic.
7 It is heated in lime kilns to make calcium oxide or **quicklime**:

$$CaCO_3\,(s) \longrightarrow CaO\,(s) + CO_2\,(g)$$

Quicklime can then be turned into calcium hydroxide or **slaked lime** by adding water. The lumps of calcium oxide get hot, swell and crumble to a white powder:

$$CaO\,(s) + H_2O\,(l) \longrightarrow Ca(OH)_2\,(s)$$

Like limestone, quicklime and slaked lime are bases, and cheap, so they too are used to neutralize acidic soil.

Limestone is hard and strong. St. Paul's Cathedral in London is built from it. It was completed in 1708.

Marble is very hard, and can be polished to a very smooth finish. It has often been used for statues.

The properties of carbonates

1 They are insoluble in water, except for sodium, potassium and ammonium carbonates.
2 They react with acids to form a salt, water and carbon dioxide.
3 Most of them break down on heating, to carbon dioxide and an oxide. You saw opposite how calcium carbonate breaks down. But sodium and potassium carbonates are not affected by heat. (The more reactive the metal, the more stable its compounds.)

Carbon dioxide

Air contains a small amount of carbon dioxide gas. Some of this is produced by **respiration**, which goes on in all living things (page 172). And some is produced by the burning of **fuels** – coal, oil, gas and wood. All these fuels are made of carbon compounds.

Properties of carbon dioxide   These are the main ones:

1 It is a colourless gas, with no smell.
2 It is much heavier than air.
3 When it is cooled to $-78\ °C$, it turns straight into a solid. Solid carbon dioxide is called **dry ice**. It sublimes when it is heated.
4 Carbon dioxide does not usually support combustion. That is why it is used in fire extinguishers.
5 It is slightly soluble in water, forming an acidic solution called carbonic acid (page 175). This is a very weak acid.

Carbon dioxide fire extinguishers are used mainly for electrical fires.

Test for carbon dioxide   For this, **lime water** is needed. Lime water is a solution of calcium hydroxide in water. Carbon dioxide makes it go milky, because a fine white precipitate forms. But when more carbon dioxide is bubbled through, the precipitate disappears. The precipitate is calcium carbonate, from this reaction:

$$Ca(OH)_2\ (aq) + CO_2\ (g) \longrightarrow CaCO_3\ (s) + H_2O\ (l)$$

It disappears again because it reacts with more carbon dioxide to form calcium hydrogen carbonate, which is soluble:

$$CaCO_3\ (s) + CO_2\ (g) + H_2O\ (l) \longrightarrow Ca(HCO_3)_2\ (aq)$$

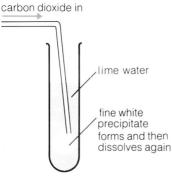

carbon dioxide in

lime water

fine white precipitate forms and then dissolves again

The test for carbon dioxide.

Uses of carbon dioxide   These are the main ones:

1 It is used in fire extinguishers.
2 It is put in drinks like cola and lemonade, to make them fizzy. It is only slightly soluble, so it is bubbled into these drinks under pressure, to make more dissolve. When the bottles are opened it escapes again, and that causes the 'fizz'.
3 Solid carbon dioxide (dry ice) is used to keep food frozen.

Questions

1 Name and describe the allotropes of carbon.
2 Name three different forms of calcium carbonate.
3 Write down two names for: **a** CaO **b** $Ca(OH)_2$
  How is the second compound made from the first?
4 Why are calcium carbonate, calcium oxide, and calcium hydroxide sometimes spread on fields?

5 Write a balanced equation for the decomposition of magnesium carbonate on heating.
6 Carbon dioxide *sublimes*. What does that mean?
7 Describe the test for carbon dioxide.
8 Write down three uses of carbon dioxide, and explain what makes it suitable for each use.

12.2 Carbon in living things

All living things contain carbon compounds:

Grass is about 4% carbon by weight.

The hard shell of this insect is almost 40% carbon.

Humans contain hundreds of different carbon compounds. This baby is about 20% carbon.

From air to living things

Carbon dioxide is the source of all the carbon compounds in living things. This is what happens:

1 First, plants take in carbon dioxide from the air, and water from the soil, to make a sugar called **glucose**. This process is called **photosynthesis**. It takes place in the plant's leaves. It needs both heat and light energy from sunlight. Chlorophyll, the green substance in leaves, acts as a catalyst for the reaction:

$$\text{carbon dioxide} + \text{water} \xrightarrow[\text{chlorophyll}]{\text{sunlight}} \text{glucose} + \text{oxygen}$$

$$6CO_2\,(g) + 6H_2O\,(l) \longrightarrow C_6H_{12}O_6\,(s) + 6O_2\,(g)$$

2 Inside the plant, the glucose is converted to **starch**, **proteins** and other carbon compounds that are needed for roots, stems and leaves. For proteins the plant also needs nitrogen, and small amounts of sulphur, which it obtains from the soil.

3 Animals obtain carbon compounds by eating plants. And humans eat both animals and plants.

Now look again at the photosynthesis reaction. It produces **oxygen** as well as glucose. The oxygen goes off into the air. Every year plants put millions of tonnes of oxygen into the air. At the same time, we take millions of tonnes of it from the air for respiration (see below) and for burning fuels.

Without photosynthesis our supply of oxygen would run out.

From living things to air

Carbon dioxide gets back to the air in these ways:

1 Plants and animals give out carbon dioxide during **respiration**. Respiration is the reaction that gives living things the energy they need. It takes place in their cells:

$$\text{glucose} + \text{oxygen} \longrightarrow \text{carbon dioxide} + \text{water} + \text{energy}$$

Note that this is the opposite reaction to photosynthesis.

Plants take in carbon dioxide and give out oxygen through **stomata** in their leaves. This is a stoma in a wheat leaf, enlarged 5000 times.

Tropical forests and jungles play a large part in making sure we do not run out of oxygen.

During photosynthesis heat and light energy are taken in and stored as chemical energy in glucose. Respiration releases the stored energy from the glucose again.

2 Bacteria feed on dead plants and animals, and produce carbon dioxide at the same time.

3 Wood, coal, oil, and natural gas are all made of carbon compounds. (They are all formed from the remains of living things.) They all produce carbon dioxide when they burn in plenty of air.

The carbon cycle

You saw on the opposite page that carbon dioxide is removed from the air by plants, and put back again in other ways. The whole process is called the **carbon cycle**:

Breathing out carbon dioxide from respiration.

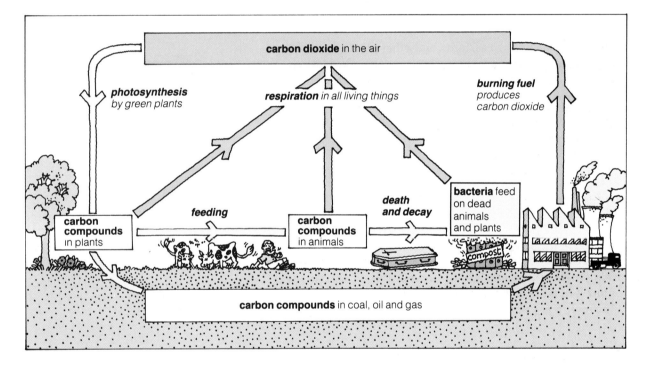

Too much carbon dioxide?   Many scientists think that we may upset the carbon cycle by burning too much fuel. This could put too much carbon dioxide into the air. The gas might then act as a blanket around the Earth, keeping in heat. Scientists call this the **greenhouse effect**. You can find out more about it on page 190.

Questions

1 What is photosynthesis? In which part of a plant does it take place?

2 Describe how carbon dioxide from the air is converted to a protein in your body.

3 Describe three ways in which carbon dioxide is returned to the air. Give equations if you can.

4 Respiration is the opposite reaction to photosynthesis. Explain why.

12.3 The fossil fuels

Most of the energy used to cook food, drive cars, and keep us warm comes from coal, oil, and gas.

1 Coal, oil, and gas are **fossil fuels**. That means they are the remains of plants and animals that lived millions of years ago.
2 They are all made of carbon compounds.
3 They are used as fuels because they give out plenty of heat energy when they burn.
4 They produce carbon dioxide and water vapour as well as energy when they burn. For example, North Sea gas (methane) burns like this:

$$CH_4 (g) + 2O_2 (g) \longrightarrow CO_2 (g) + 2H_2O (g) + energy$$

Coal

Coal is the remains of trees, ferns, and other plants that lived as much as three million years ago. These were crushed into the Earth, perhaps by earthquakes or volcanic eruptions. They were pressed down by layers of earth and rock. They slowly decayed into coal.

Around 50 million tonnes of coal are mined each year in Britain. The table shows that most of it is burned in power-stations, to make electricity. The electricity is then used for heating, lighting, and driving machinery.

Coal used for	Amount (million tonnes)
Power-stations	30
Coke ovens	5
Industry	5
Home heating	5
Exports	3
Offices, hospitals etc.	2

Oil and gas

Oil and gas are the remains of millions of tiny plants and animals that lived in the sea. When they died, their bodies sank to the sea-bed and were covered by silt. Bacteria attacked the dead remains, turning them into oil and gas. Meanwhile the silt was slowly compressed into rock. The oil and gas seeped into the porous parts of the rock, and got trapped like water in a sponge.

Where does electricity fit in?

Electricity is not itself a fuel. But it is made from fuels:

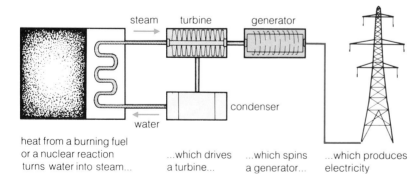

heat from a burning fuel or a nuclear reaction turns water into steam...   ...which drives a turbine...   ...which spins a generator...   ...which produces electricity

Although oil was formed under the sea, many oil wells are on dry land. Millions of years ago, movement of the Earth's crust forced some sea-beds upwards to form land.

From coal to coke

Some coal produces a lot of smoke and harmful tarry fumes when it burns. It must be made 'smokeless' before it can be used in factories and homes. This is done by heating the coal in the absence of air. **Coal gas** and **coal tar** are driven from the coal, leaving a fuel called **coke** behind.

They are not just fuels . . .

Crude oil is a mixture of many carbon compounds. The first step is to separate it into groups of similar compounds (page 193). Some of these are used as fuels. But others are turned into plastics, man-made fibres, drugs, pesticides, insecticides and detergents.

Most coal is used as a fuel. But coke is used in the blast furnace to reduce iron ore to iron. And coal tar is used to make things like inks, detergents, and insecticides. (In fact anything made from oil can also be made from coal. But at the moment it would cost more.)

How long will they last?

This diagram shows how much we depend on coal, oil and gas.

Smokeless fuels. These burn much more cleanly than ordinary coal. But they still produce sulphur dioxide, because they contain sulphur compounds.

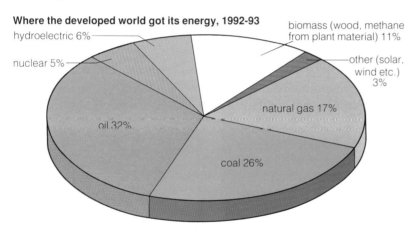

Where the developed world got its energy, 1992-93

hydroelectric 6%
nuclear 5%
biomass (wood, methane from plant material) 11%
other (solar, wind etc.) 3%
natural gas 17%
oil 32%
coal 26%

But this state of affairs cannot last. Coal, oil and gas are **non-renewable** sources of energy. They will run out eventually — maybe 20 or 50 years from now. The uranium for nuclear power-stations will also get used up. So we will depend more and more on **renewable** sources of energy such as wind and solar power.

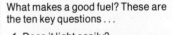

What makes a good fuel? These are the ten key questions . . .

1 Does it light easily?
2 Does it burn steadily?
3 Does it give out much energy?
4 Does it pollute the air?
5 Does it leave much ash behind?
6 Is it plentiful?
7 Does it cost a lot?
8 Is it easy and safe to store?
9 Is it easy and safe to transport?
10 Might it be better used for something else?

Questions

1 What does the term *fossil fuel* mean?
2 Describe how coal was formed.
3 What is the main use of coal at present?
4 Oil and gas are usually found together. Why?
5 How is coke made from coal?

6 Name three products made from coal tar.
7 Why do we need to find other forms of energy?
8 Using the list of ten questions above, decide whether: **a** paper **b** magnesium ribbon would make a good alternative fuel.

12.4 Fossil fuels and the environment

We have been burning up fossil fuels so fast over the last 150 years that our environment is in real danger.

Carbon dioxide and the greenhouse effect

As you saw on page 190, life could not go on without carbon dioxide. Plants take it from the air for photosynthesis, and living things give it back through respiration. Together, these processes can keep the amount of carbon dioxide in air in balance.

But the burning of fossil fuels also produces carbon dioxide. It is given off from car engines, home central heating systems, gas cookers, power-stations and factories. At the same time, the tropical rain forests which use it up are being destroyed. So it is no longer in balance.

Many scientists accept that the extra carbon dioxide is acting as a blanket round the Earth, keeping in heat. They call this the **greenhouse effect**. They think it will cause **global warming**, the ice around the Poles will melt, and there will be huge floods.
And in fact over the last hundred years, the average air temperature around the Earth has already increased by half a degree.

Some unpleasant things being added to air.

Pollution from the fossil fuels

The fossil fuels (and fuels like petrol made from them) also cause other problems when they burn. They produce these pollutants.

Carbon monoxide   When coal, gas and oil burn in too little air, carbon monoxide is produced instead of carbon dioxide. This gas has the formula CO. It is colourless, insoluble in water, and has no smell. But more importantly, it is poisonous. It reacts with the **haemoglobin** in your blood and stops it carrying oxygen to your body cells. So you could die from oxygen starvation.

About 85% of the carbon monoxide in the air in Britain comes from the exhausts of cars, trucks and buses.

Hydrocarbons   Not all the petrol in car engines gets burned. Some escapes in the exhaust. Some also escapes when tanks are filled at petrol stations. The unburned petrol contains benzene, which is known to cause cancer, and other hydrocarbons that cause liver damage.

Sulphur dioxide   Coal, oil and gas all contain sulphur compounds. If these are not removed, they form sulphur dioxide when the fuel is burned. This gas attacks the lungs and can affect people with asthma very badly. It also dissolves in rainwater to form sulphurous acid (page 177). This **acid rain** damages buildings, trees and plants, and kills living things in rivers.

Nearly three-quarters of the sulphur dioxide in the air in Britain comes from power-stations. Some also comes from diesel engines.

The air over the British Isles contains about 8.5 million tonnes of carbon monoxide, mostly from car engines. Luckily it is not concentrated enough to kill us immediately! However it is particularly harmful to people with heart disease.

Nitrogen oxides These are an indirect result of burning fossil fuels. Inside car engines and power-station furnaces, the air gets so hot that its nitrogen and oxygen react together, forming oxides of nitrogen. Like sulphur dioxide, these attack the lungs and cause acid rain.

About 45% of the nitrogen oxides in our air in Britain comes from vehicle exhausts, and about 35% from power-stations.

Ozone This is also a by-product of burning fossil fuels. It is formed in heavy traffic in hot weather, when sunlight causes the nitrogen oxides and hydrocarbons from car exhausts to react together.

Up in the ozone layer, ozone protects us from the damaging rays of the Sun. But at ground level it can cause sore eyes, noses and throats, headaches, breathing problems and lung damage.

Tiny solid particles These are mainly carbon particles that come from burning coal. If lead-free petrol is not used, exhaust fumes will also contain particles of lead from a compound called **tetra-ethyl lead** that is added to petrol to make it burn better.

The carbon particles are coated with harmful hydrocarbons and other compounds, and can get deep into your lungs. They also damage trees and plants, and make buildings black and grimy. Lead causes brain damage, particularly in children.

Stonework worn away by acid rain.

What can be done?

Several things are being done to cut down pollution from fossil fuels:

1 The exhausts of all new cars are fitted with **catalytic converters**, where harmful gases react to form harmless ones. (But carbon dioxide is not affected.)
2 All new cars run on lead-free petrol.
3 Manufacturers are looking at ways to make cars engines more efficient so that they use less petrol, and at fuels that could be used instead of petrol.
4 Coal is turned into **smokeless fuel** for using in homes (page 189).
5 The European Union has ordered all power-stations to cut the sulphur dioxide they give out by 60%, by the year 2003.
6 Governments are looking at ways to make homes and factories more energy efficient, so that we burn less fuel, not more.

But there is a lot more that could be done. For example, if there was cheap and efficient public transport, people would use their cars less.

A more environmentally-friendly way to travel?

Questions

1 Explain why carbon dioxide, which is essential to life, can also be a danger.
2 a What is carbon monoxide and how is it formed?
   b Why is carbon monoxide so dangerous?

3 Write a list of all the pollutants in the exhaust from car engines.
4 Write a list of things *you* could do to help cut down on the amount of fossil fuel being burned.

12.5 Chemicals from oil (I)

Oil is the remains of tiny sea animals and plants that lived millions of years ago. It is a mixture of hundreds of different carbon compounds. Their molecules are chains or rings of carbon atoms with other atoms bonded on. For example:

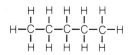

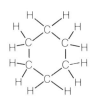

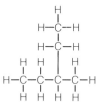

This molecule has a chain of 5 carbon atoms.

Here a chain of 6 carbon atoms has formed a ring.

Here 6 carbon atoms form a **branched** chain.

In these examples, the molecules contain only carbon and hydrogen atoms. So the compounds are called **hydrocarbons**.
A hydrocarbon contains only hydrogen and carbon atoms.
Most of the compounds in oil are hydrocarbons.

Using the compounds from oil

The compounds obtained from oil have thousands of different uses. That is why countries with a lot of oil to sell tend to get rich!

Some of the compounds are used as fuels for cars, trucks, planes and ships.

Some are the starting point for detergents, drugs, dyes, paints and cosmetics.

Some are the starting point for nylon, PVC and other plastics. Most plastics start off as oil.

Getting oil from the Earth

Oil is usually found about 3 or 4 kilometres below ground, trapped in rocks like water in a sponge. In Britain, oil-bearing rocks lie beneath the North Sea. To get at the oil, a hole is bored in the sea-bed, using a giant drill.
The drill is hung from a **drilling platform** like the one on the right. The hole is lined with steel. Then the oil is pumped up through it, and into pipes that run along the sea-bed. These pipes carry it to a tanker terminal, or all the way ashore to an **oil refinery**.

An oil platform in the North Sea.

Refining the oil

Some oil compounds have small molecules with only a few carbon atoms. But some have as many as 50 carbon atoms. To make the best use of it, the oil is separated into groups of compounds that are close in size. This is called **refining** the oil.

Refining is carried out by **fractional distillation** in a tall tower, which is kept very hot at the base, and cooler towards the top. Crude oil is pumped in at the base, and the compounds start to boil off. Those with the smallest molecules have the lowest boiling points. So they boil off first and rise to the top of the tower. Others rise only part of the way, depending on their boiling points, and then condense.

Oil tanker jetty.

cool (25 °C)

fractions out

crude oil in

very hot (over 400 °C)

Name of fraction	Number of carbon atoms	What the fraction is used for
Gas	C_1 to C_4	Separated into the fuels methane, ethane, propane and butane.
Gasoline	C_5 to C_6	Blended with other fractions to make petrol.
Naphtha	C_6 to C_{10}	Starting point for many chemicals and plastics.
Kerosene	C_{10} to C_{16}	Jet fuel. Detergents.
Diesel oil	C_{16} to C_{20}	Fuel for central heating.
Lubricating oil	C_{20} to C_{30}	Oil for cars and other machines.
Fuel oil	C_{30} to C_{40}	Fuel for power-stations and ships.
Paraffin waxes	C_{40} to C_{50}	Candles, polish, waxed papers, waterproofing, grease.
Bitumen	C_{50} upwards	Pitch for roads and roofs.

boiling points and viscosity increase

As molecules get larger, the fractions get less runny, or more **viscous**: from gas at the top of the tower to solid at the bottom. They also get less **flammable**. Grease burns less easily than petrol, for example. So the lower fractions are not used as fuels.
Once the fractions are removed, they need further treatment before they can be used. You can find out about this on the next page.

Questions

1 Copy and complete: Oil is the remains of tiny prehistoric . . . . and . . . . It is a . . . . . of compounds. These are mostly . . . . which means they contain just . . . . . and . . . .

2 Oil is *refined* before use. What does that mean?

3 Name the fraction where the compounds have the:
   a smallest molecules; b lowest boiling points.

4 a The bitumen fraction is the least viscous and has the highest boiling point. Why is this?
   b Why wouldn't you use bitumen as a fuel?

193

12.6 Chemicals from oil (II)

You saw in the last Unit how crude oil is separated into **fractions** by fractional distillation. But that's not the end of the story. The fractions all need further treatment before they can be used.

1 They contain impurities. These are mainly sulphur compounds. If left in fuel, they will burn to form poisonous sulphur dioxide gas.
2 Some fractions are separated into single compounds or smaller groups of compounds. For example, the gas fraction is separated into methane, ethane, propane and butane.
3 Part of a fraction may be **cracked**. Cracking breaks down molecules into smaller ones.

This site removes sulphur impurities.

Cracking

Cracking is very important for two reasons:

1 It lets you turn long-chain molecules into more useful shorter ones. For example you may have plenty of naphtha, but not enough gasoline for making petrol. So you could crack some naphtha to get molecules the right size for petrol.
2 It *always* produces short-chain compounds like ethene and propene, which have a double bond between carbon atoms. The double bond makes these compounds **reactive**. So they can be used to make plastics and other substances.

This is the kind of reaction that takes place when you crack naphtha:

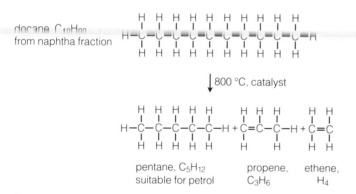

This is the cracking plant at Esso's Fawley refinery.

decane, $C_{10}H_{22}$ from naphtha fraction

↓ 800 °C, catalyst

pentane, C_5H_{12} suitable for petrol

propene, C_3H_6

ethene, H_4

When a catalyst is used, the process is called **catalytic cracking**. Now look at the reaction below. It shows how you can even crack ethane, already a small molecule, to give ethene, by mixing it with steam at high temperatures:

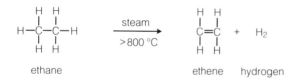

ethane

steam
>800 °C

ethene   hydrogen

In fact you can crack any hydrocarbon molecule that has two or more carbon atoms with single bonds between them.

From cracking to plastics

The two cracking reactions shown opposite produced ethene.
Look how ethene is used to make plastics:

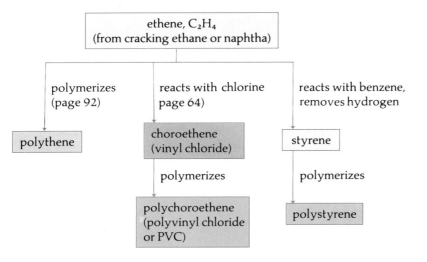

Other compounds from the cracking of naphtha and gas oil are also
polymerized to make plastics. For example:

propene (C_3H_6) polymerizes to give **polypropene**

butene (C_4H_8) polymerizes to give **synthetic rubber**.

Reforming

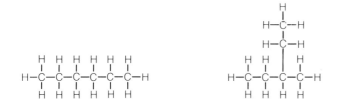

Look at these hydrocarbons. Both have the same formula, C_6H_{12}.
Both burn well, and are used in petrol. But the one with the straight
chain will **ignite** or catch fire more easily. In fact it may ignite a little
too soon in the hot cylinder of a car engine, so the engine will not
work as efficiently as it should.
For this reason, high-quality petrol contains mainly hydrocarbons
with branched chains. These are made by taking molecules with
straight chains from the gasoline and naphtha fractions, and making
the chains branch. This process is called **reforming**.

Only one of the products refined from oil.

Questions

1 What is *cracking*? Why is it so important?
2 **a** A straight-chain hydrocarbon has the formula
   C_5H_{12}. Draw a diagram of one of its molecules.
   b Now show what might happen when it is cracked.

3 **a** What does *reforming* mean?
   b Why are compounds reformed to make petrol?
4 Take the molecule you drew for question 2 and show
   how it might look when reformed.

12.7 Alkanes and alkenes

The carbon compounds in living things, and in coal, oil and gas (which were once living things) are called **organic compounds**. They can be divided into families, and here we will look at two of those families, the **alkanes** and **alkenes**. Both are hydrocarbons – they contain only carbon and hydrogen atoms.

Alkanes

This table shows the first four members of the alkane family:

Name	Methane	Ethane	Propane	Butane
Formula	CH_4	C_2H_6	C_3H_8	C_4H_{10}
Structure of molecule	$H-\underset{\underset{H}{|}}{\overset{\overset{H}{|}}{C}}-H$	$H-\underset{\underset{H}{|}}{\overset{\overset{H}{|}}{C}}-\underset{\underset{H}{|}}{\overset{\overset{H}{|}}{C}}-H$	$H-\underset{\underset{H}{|}}{\overset{\overset{H}{|}}{C}}-\underset{\underset{H}{|}}{\overset{\overset{H}{|}}{C}}-\underset{\underset{H}{|}}{\overset{\overset{H}{|}}{C}}-H$	$H-\underset{\underset{H}{|}}{\overset{\overset{H}{|}}{C}}-\underset{\underset{H}{|}}{\overset{\overset{H}{|}}{C}}-\underset{\underset{H}{|}}{\overset{\overset{H}{|}}{C}}-\underset{\underset{H}{|}}{\overset{\overset{H}{|}}{C}}-H$
Number of carbon atoms in chain	1	2	3	4
Boiling point	$-164\,°C$	$-87\,°C$	$-42\,°C$	$-0.5\,°C$

boiling point increases with chain length

One more carbon atom is added each time. It could reach thousands! The general formula for the family is C_nH_{2n+2} where n is a number. Can you see how this formula fits the alkanes in the table?

Points to note Note these things about the alkanes:

1 They are found in oil and natural gas. Natural gas is mostly methane, with small amounts of ethane, propane and butane. The mixture of compounds in oil includes alkanes with chains of up to 50 carbon atoms.
2 The longer the chain, the higher the boiling point. So the first four alkanes are gases at room temperature, the next twelve are liquids, and the rest are solids.
3 In an alkane molecule, each carbon atom forms four single covalent bonds. The bonding in ethane is shown on the right.
4 Alkanes are unreactive. They are not affected by acids or alkalis.
5 But they burn well in a good supply of oxygen, forming carbon dioxide and water vapour, and giving out plenty of heat. For example, propane burns like this:

$$C_3H_8\,(g) + 5O_2\,(g) \longrightarrow 3CO_2\,(g) + 4H_2O\,(g) + heat$$

So alkanes are used as fuels. Natural gas (North Sea gas) is used for cooking and heating in homes. Propane and butane are used as camping gas, Calor gas is mainly butane. Petrol contains butane and liquid alkanes, as well as many other hydrocarbons.

one shared pair of electrons (a single bond)

The bonding in ethane.

Camping gas is butane.

196

Alkenes

Here are the first two members of the alkene family:

Name	Ethene	Propene
Formula	C_2H_4	C_3H_6
Structure of molecule	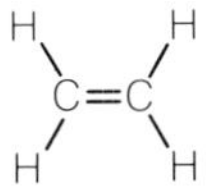 H\C=C/H / \ H H	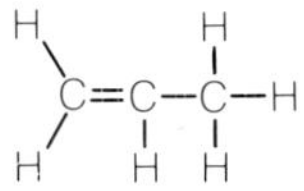 H\C=C−C−H / H H

The general formula for the alkene family is C_nH_{2n}. Can you see why?

Points to note   Note these things about the alkenes:

1 They are made from alkanes by **cracking** (page 92).
2 An alkene molecule contains a **double bond** between carbon atoms. (Alkanes have only single bonds.) The bonding in ethene is shown on the right.
3 Alkenes are much more reactive than alkanes. This is because the double bond can break to form single bonds and add on other atoms. For example, ethene reacts with hydrogen like this:

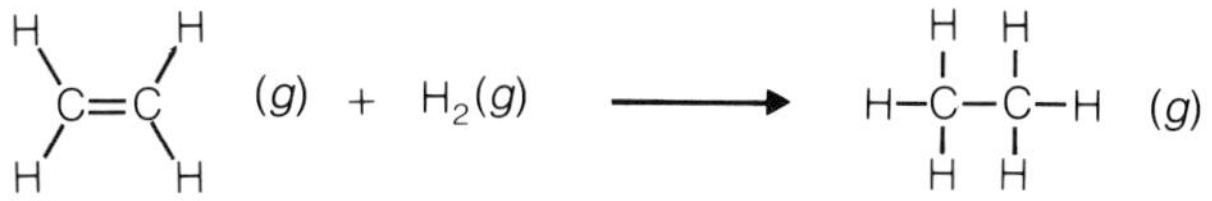

ethene (g) + $H_2(g)$ ⟶ ethane (g)

This is called an **addition reaction**.
4 In the same way alkene molecules can also add on to each other, to make long-chain molecules called **polymers**. See page 92.
6 Because the double bond allows them to add on more atoms, alkenes are said to be **unsaturated**. (Alkanes are saturated.)

Tests for unsaturation   There are two ways you could test a hydrocarbon, to see whether it is an alkane or an alkene:

1 Add bromine water. This is an orange solution of bromine in water. It turns colourless in the presence of an alkene because the bromine adds on to the alkene, giving a colourless compound:

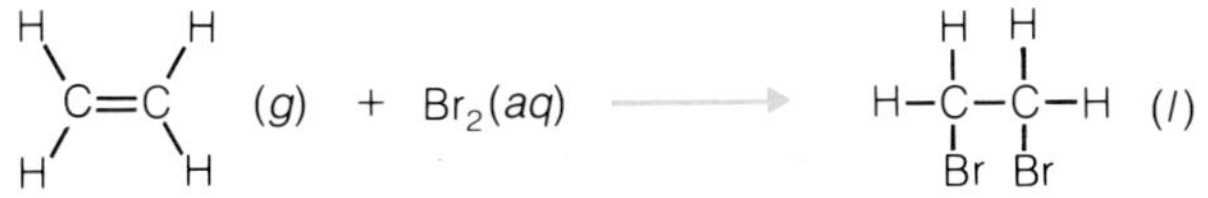

ethene (g) + $Br_2(aq)$ ⟶ 1,2-dibromoethane (l)

2 Add acidified potassium manganate(VII) solution. This is purple, but turns colourless if an alkene is present.

Unsaturated hydrocarbons don't burn cleanly.

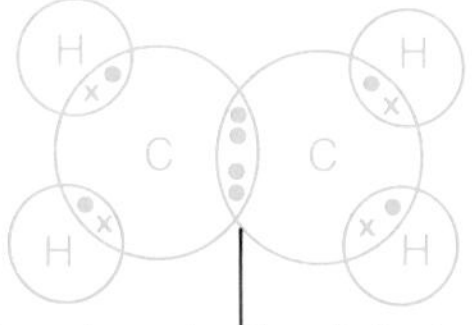

two shared pairs of electrons (a double bond)

The bonding in ethene.

Questions

1 Will alkanes conduct electricity? Explain.
2 Draw a diagram to show the *bonding* in methane.
3 What is the difference between an *ethane* molecule and an *ethene* molecule?
4 Alkenes are *unsaturated*. What does that mean? What effect does it have?
5 Write an equation to show how propene reacts with bromine water. What change would you see?

12.8 Alcohols

Now we look at another family of organic compounds: **alcohols**. The two simplest alcohols are methanol and ethanol.

methanol, CH_3OH ethanol, C_2H_5OH

Both compounds have an $-OH$ group bonded to a carbon atom. **Alcohols are organic compounds that contain an $-OH$ group.**

Methylated spirit.

Ethanol

Ethanol is the best-known alcohol. It is often just called 'alcohol'. It is a good solvent, dissolving many substances that are insoluble in water. It also evaporates quickly. So it is used in glues, paints, varnishes, printing inks, deodorants, colognes and aftershaves. It is also used as a raw material for making other substances such as synthetic rubbers and flavourings.
Ethanol is well known for another reason too — it makes people drunk. Beer, wine, and other alcoholic drinks all contain ethanol.

The properties of ethanol

1 It is a clear, colourless liquid, that boils at 78 °C.
2 The pure liquid is dangerous to drink, and even dilute solutions of it affect the body. At first, a dilute solution makes you feel relaxed, but too much causes drunkenness, with headache, dizziness and vomiting. Over time it can ruin your liver.
3 It burns well in air or oxygen, giving out heat:

$$C_2H_5OH\,(l) + 3O_2\,(g) \longrightarrow 2CO_2\,(g) + 3H_2O\,(g)$$

As methylated spirits, it is often burned as a fuel, in **spirit lamps**. These are used by jewellers, and sometimes by cooks.
4 Ethanol is slowly oxidized to **ethanoic acid**, by bacteria in the air:

$$C_2H_5OH\,(l) + O_2\,(g) \longrightarrow CH_3COOH\,(aq) + H_2O\,(l)$$

This reaction causes wine to go 'sour' when left open in air.

Ethanol is used in alcoholic drinks, aftershave and perfume.

Making ethanol for industry

The solvent ethanol is made by mixing ethene and steam, and passing the mixture over a catalyst. This addition reaction takes place:

ethene   ethanol

The product is in fact a solution of ethanol in water. Most of the water is then removed by fractional distillation (page 22).

Unfortunately, some people become addicted to ethanol. They turn into alcoholics.

Making ethanol by fermentation

The ethanol in alcoholic drinks is made by **fermentation**. During fermentation, glucose from fruit, vegetables or cereals such as barley, is turned into ethanol by natural catalysts called **enzymes**. The enzymes are contained in **yeast**.

$$\text{glucose} \xrightarrow[\text{yeast}]{\text{enzymes in}} \text{ethanol} + \text{carbon dioxide}$$
$$C_6H_{12}O_6 \, (s) \longrightarrow 2C_2H_5OH \, (l) + 2CO_2 \, (g)$$

Making beer   Beer is made from **barley** by fermentation. First barley grains are soaked in hot water to start them germinating. The germinating barley, called **malt**, is left in air, and enzymes in it start breaking down the starch into glucose.
Next the malt is dried, crushed and mixed well with water. The barley husks are filtered off, leaving a solution of glucose and enzymes. This is boiled to stop the enzymes working, and hops are added for flavour. Then it is cooled, yeast is added, and fermentation begins.

During fermentation, the yeast converts the glucose to ethanol. When enough ethanol has formed, the yeast is removed and is used again for the next batch of beer.

Making wine   To make white wine, grapes are crushed and then the skins are filtered off, leaving the juice behind. Yeast is added and fermentation begins. It carries on until either the yeast is removed, or all the glucose in the juice is used up.
For red wine, the yeast is added to the crushed red grapes before the skins are filtered off, so that the juice has time to extract the red colour from them.

How alcoholic is it?

How much ethanol is in an alcoholic drink? A pint of strong beer contains about twice as much as a pint of ordinary beer. Like beer, whisky is made from barley. But it is more 'alcoholic' because the fermented liquid is distilled to make the ethanol more concentrated.

The drinks below all contain about the same amount of ethanol:

A starting point for ethanol!

Beer fermenting. The froth is caused by escaping carbon dioxide. Fermentation is carried out at 18–20 °C. At lower temperatures the reaction is too slow. At higher temperatures the yeast may get killed.

  = = =

A half pint of ordinary beer   A small glass of sherry   A small whisky (or gin or vodka)   A glass of wine

Questions

1 List three uses of ethanol.
2 Why does wine go sour when left in air?
3 Describe what happens during fermentation.
4 Why does the temperature need to be carefully controlled during fermentation?
5 Why is whisky more 'alcoholic' than beer?

Questions on Section 12

1 a Copy this diagram and complete it by writing in:
  i the common names;   **ii** the chemical formulae.

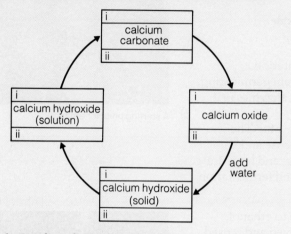

b Beside each arrow say how the change would be carried out. One example is shown.
c Give three reasons why limestone is an important raw material.

2 Sodium hydrogen carbonate is often called sodium bicarbonate or bicarbonate of soda. Its formula is $NaHCO_3$. When heated it gives off a gas. It is used in baking soda, baking powder and indigestion tablets.
a Name the gas given off when bicarbonate of soda is heated. (Hint: it is also given off when limestone is heated!)
b Explain why the compound is used for baking.
c Why is sodium *carbonate* no use for baking?
d Hydrogen carbonates react like carbonates with acids. What products will be formed when sodium hydrogen carbonate reacts with hydrochloric acid?
e Write a balanced equation for this reaction.
f Explain why sodium hydrogen carbonate is used in indigestion tablets.

3 Below are ten descriptions of a gas. Which of these describe the gas carbon dioxide?
a Colourless.
b Given out during photosynthesis.
c Turns lime water milky.
d Burns in air.
e Insoluble in water.
f Heavier than air.
g Has the formula CO_2.
h Used up in burning carbon compounds.
i Has no smell.
j Is found in air.
k A product of burning fossil fuels.
l Forms a weak acid when it dissolves in water.

4 Put these sentences in the correct order to make a description of the carbon cycle.
a In the reaction, carbon dioxide and water are given out.
b The carbon dioxide and water combine to form sugars.
c The reaction also gives out a lot of energy, which helps to keep the animals warm.
d All green plants take in carbon dioxide and water.
e Animals eat plants for food.
f Sunlight provides the energy for this reaction.
g The sugar from the food reacts with oxygen in the cells of the animals.

5 Catalytic converters are fitted to all new cars to reduce pollution caused by hydrocarbons, carbon monoxide and oxides of nitrogen in the exhaust gas.
a What are the main dangers associated with each of these pollutants?
b What is meant by a *catalytic* reaction?
c In one of the catalytic reactions nitrogen monoxide (NO) reacts with carbon monoxide to form nitrogen and carbon dioxide. Write a balanced equation for this reaction.
d What environmental problem is *not* solved by the use of catalytic converters?
e Why is the level of ozone at ground level increasing? Is this a good or a bad thing? Why?

6 Use the information on pages 196 and 197 to answer these questions about the alkanes:
a Which two elements do alkanes contain?
b Which alkane is the main compound in natural gas?
c After butane, the next two alkanes in the series are *pentane* and *hexane*. How many carbon atoms would you expect to find in a molecule of:
  i pentane?   **ii** hexane?
d Write down the formulae for pentane and hexane.
e Draw a molecule of each substance.
f Is pentane a solid, liquid or a gas at room temperature?
g Suggest a value for the boiling point of pentane, and explain your answer.
h Would pentane react with bromine water? Explain.
i Alkanes burn in a good supply of oxygen. Name the gases formed when they burn.
j Write a balanced equation for the burning of pentane in oxygen.
k Butane, pentane and hexane have different *chain lengths*. Explain what that means.
l Why are butane, pentane, and hexane all useful as fuels?

7 Ethanol is a member of a family of compounds called alcohols.

a Write the chemical formula of ethanol.

b How does a molecule of ethanol differ from a molecule of ethane?

c Ethanol is one of the products of the fermentation of sugar. The diagram shows apparatus which could be used to study the fermentation.

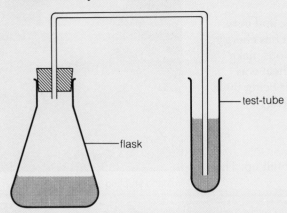

What should go in the flask?

d Which of these temperatures would be best for the reaction?

    i 0 °C   **ii** 10 °C   **iii** 25 °C   **iv** 55 °C

Explain your choice.

e The test-tube contains water, which prevents air reaching the flask. Explain why this is important.

f Which gas is released during the fermentation?

g Complete the equation for the reaction:

    $C_6H_{12}O_6 \longrightarrow$

    (sugar)

h How long would you expect the reaction to take?

    i 5 minutes   **ii** 5 hours   **iii** 5 days   **iv** 5 months

i What process would you use to separate a reasonably pure sample of ethanol from the mixture?

8 Propene is one of the many important hydrocarbons made from oil. Like *propane*, it is made up of molecules which contain three carbon atoms. Like *ethene*, it has a double bond.

a Draw a molecule of propene.

b How does it differ from a molecule of propane?

c To which group of hydrocarbons does:

    i propane   **ii** propene   belong?

d Write formulae for propane and propene.

e Which of the two is a *saturated* hydrocarbon?

f **i** Explain why propene reacts immediately with bromine water, while propane does not.

    ii What would you *see* in the reaction?

g Name another reagent that would react immediately with propene but not with propane.

h Propene is obtained by breaking down longer-chain hydrocarbons. What is this process called?

i Propene is the monomer for making an important plastic. Suggest a name for the polymer it forms.

9 Crude oil is a mixture of hydrocarbons, each with a different boiling point.

a It is an important raw material. Why?

b How is the mixture separated?

c A simple separation can be carried out in the laboratory, using this apparatus:

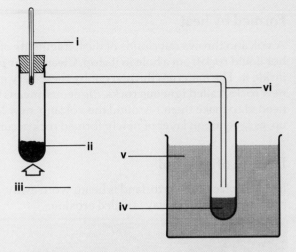

Copy the diagram and label it using these labels:
mineral wool soaked in crude oil   *delivery tube*   *thermometer*   *heat*   *water*   *fraction*

d **i** What is the purpose of the mineral wool?

    ii What is the purpose of the water?

    iii Why is the thermometer placed where it is?

10 In the experiment in question 9, a crude oil sample was separated into four fractions. These were collected in the temperature ranges shown below:

Fraction	Temperature range/°C
A	25–70
B	70–115
C	115–200
D	200–380

Which fraction:

a has the lowest range of boiling points?

b burns most readily?

c has molecules with the longest chains?

11 a Which of these molecules could be used as monomers for making plastics? Explain your choice.

ethene, $CH_2 = CH_2$ ethanol, C_2H_5OH
propane, C_3H_8 styrene, $C_6H_5CH = CH_2$
chloropropene, $CH_3CH = CHCl$

b Suggest a name for each polymer obtained.

c The polymers obtained are all thermoplastics. What special property do thermoplastics have?

d Name one thermosetting plastic. What special property do thermosetting plastics have?

e What problems can the use of plastics cause?

13.1 The changing face of the Earth

The Earth's surface features are formed and changed by many different processes.

Formed by heat

A volcano throws out clouds of dust, fragments of rock, and very hot liquid rock from a hole in its top. Clearly, our planet has energy inside it. The materials thrown out by these fiery eruptions make new rocks, called **igneous rocks**. ('Igneous' means that heat was needed to make them.) Around the volcano, new land is being built up, as layer upon layer of newly formed rock accumulates.

Volcano erupting.

Erosion of the land

In many places on Earth, land is being worn away, not built up. The process of wearing away is called **erosion**.

The Grand Canyon, Arizona, USA.

In the Grand Canyon, layers of rocks have been eroded by the Colorado River. Deep in the canyon, the process is still going on. Bits of rock are being carried away by water.

Formed from sediments

The river on the right slows as it flows into the lake, making the mud and sand it carries sink to the bottom. In time, this sediment will harden to become new layers of rocks. These **sedimentary rocks** are made of recycled material from older rocks.

Most of the old rocks of the Grand Canyon are sedimentary rocks. Those at the bottom are the oldest, because they were laid down first. The sedimentary rocks have fossils of sea animals in them. But today, those layers are a long way above the sea.

Mud and sand being washed into a lake.

Changed by heat and pressure

These sedimentary rocks have been squashed, buckled, and broken by huge forces from within the Earth.

These mountains are made of metamorphic rocks, formed when sedimentary rocks were changed by heat and pressure.

In many places on Earth, layers of sedimentary rocks have been squashed, buckled, and broken. The Earth must have plenty of energy to create forces that can do this to rocks.

The mountains in the photograph above right are made of rocks that have been buckled up and heated while buried deep under other rocks. They were sedimentary rocks in the past, but have now changed into new, harder rocks called **metamorphic rocks**. 'Metamorphic' means 'changed' — either by heat, pressure from rocks above, Earth movements, or a combination of these.

Igneous rock.

Sedimentary rock.

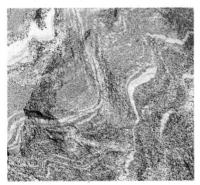

Metamorphic rock.

Questions

1 What evidence is there that the Earth has huge amounts of energy inside it?

2 In the photograph at the bottom of the opposite page, where could the mud and sand have come from?

3 If you walked down into the Grand Canyon, where would you find the oldest rocks? Why?

4 Look at the three types of rock in the photographs above.

   a Which type was formed when existing rock was changed as a result of heat or pressure?

   b Which type was formed from fragments of existing rock, deposited in layers?

13.2 Records of the Earth's past

Fossil clues

Fossils are one of the keys to the Earth's past. Layers of rock tell the story of the Earth, with fossils to put the story in order.

In the nineteenth century, William Smith noticed that each layer of sedimentary rock had its own special set of fossils. Later, other scientists drew boundaries where the fossils showed major changes. Assuming that deeper layers were older than those above, they divided time into the geological **periods** shown in the chart on the next page. The chart includes dates, but much scientific detective work was needed to estimate what those dates were.

Depth clues

In the 1830s, Charles Lyell estimated that the volcano Mount Etna in Sicily had erupted 1 km³ of rock every 100 years. Knowing the size of the volcano, he worked out that it must have started to form at least 50 000 years ago. He also looked at fossil shells in the sedimentary rock beneath the oldest lava. They were identical to sea shells found on Sicilian beaches. He concluded that the rocks below the volcano, although more than 50 000 years old, were still only from the most recent geological time period, the Quaternary.

More clues about the Earth's age came from the thickness of sedimentary layers. In the 1870s, Samuel Haughton estimated that 1 metre of sediment took about 25 000 years to form. From this, he calculated that the earliest Cambrian layers were about 600 million years old.

Mary Anning was a famous nineteenth century fossil collector. At the age of twelve, she became the first to find the fossil of a huge marine reptile, *Ichthyosaurus*.

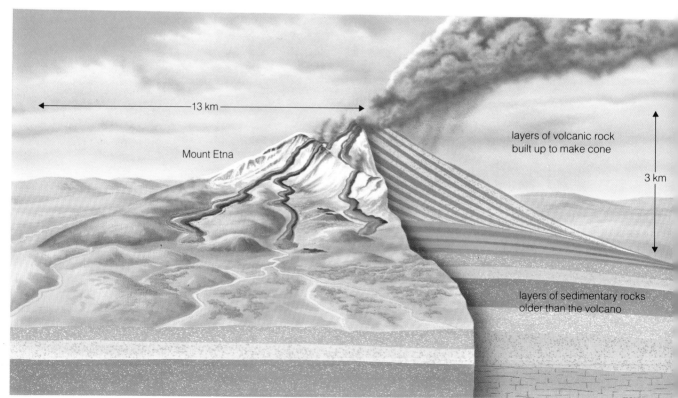

13 km

Mount Etna

layers of volcanic rock built up to make cone

3 km

layers of sedimentary rocks older than the volcano

The Earth's place in time

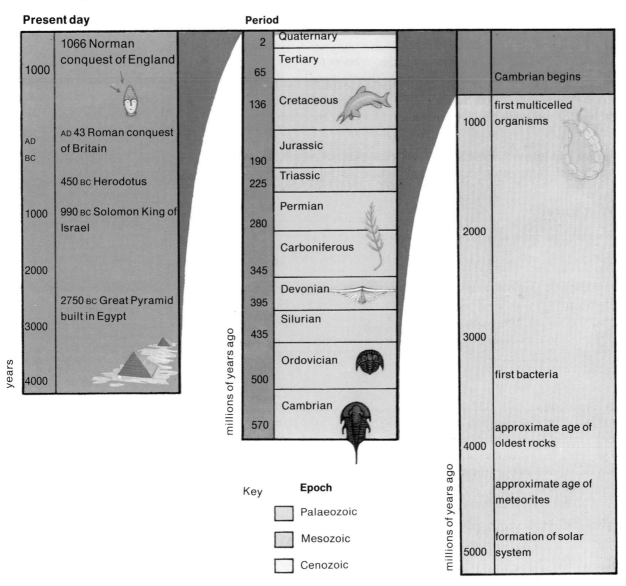

Present day

years

- 1000 — **1066 Norman conquest of England**
- AD / BC — **AD 43 Roman conquest of Britain**
- **450 BC Herodotus**
- 1000 — **990 BC Solomon King of Israel**
- 2000
- **2750 BC Great Pyramid built in Egypt**
- 3000
- 4000

Period

millions of years ago

- 2 — Quaternary
- 65 — Tertiary
- 136 — Cretaceous
- 190 — Jurassic
- 225 — Triassic
- 280 — Permian
- 345 — Carboniferous
- 395 — Devonian
- 435 — Silurian
- 500 — Ordovician
- 570 — Cambrian

Key — **Epoch**
- Palaeozoic
- Mesozoic
- Cenozoic

millions of years ago

- Cambrian begins
- 1000 — first multicelled organisms
- 2000
- 3000
- first bacteria
- 4000 — approximate age of oldest rocks
- approximate age of meteorites
- 5000 — formation of solar system

Radioactive clues

Nowadays, radioactivity can be used to date rocks. This is called **radiometric dating**. It is mainly used on igneous and metamorphic rocks. For more on the technique, see page 43.

So far, the oldest rocks found on Earth come from Greenland. They are 4000 million years old. Meteorites are up to 4600 million years old, which is probably the age of the planets, including Earth.

Questions

1 Why are the lowest layers of rock likely to contain the oldest fossils?

2 About how long ago do scientists think that the Earth was formed?

3 How can the ages of igneous and metamorphic rocks be estimated?

4 Why is it likely that the sedimentary rocks under Mount Etna are at least 50 000 years old?

13.3 The atmosphere past and present

The **atmosphere** is the layer of gas around the Earth. It may feel light, but when you lie flat out at the seaside to sunbathe, there is about 13 000 kilograms of gas pressing down on your body!

This gas layer has four parts. We live in the **troposphere**, with trips into the **stratosphere** now and again. Only a few humans have reached the **mesosphere**. The **ionosphere** is mainly charged particles.

The gas is at its most dense at sea-level, but thins out rapidly as you rise through the troposphere. It is the mixture we call **air**.

What's in air?

This pie chart shows the gases that make up air:

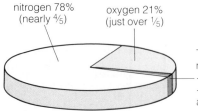

nitrogen 78% (nearly ⁴⁄₅)

oxygen 21% (just over ¹⁄₅)

The remaining 1% is...
mainly argon + a little carbon dioxide
+ a little water vapour
+ smaller amounts of helium, neon, krypton, and xenon (the other noble gases)

The composition of air is not exactly the same everywhere. It changes a little from day to day and place to place.
For example, there is more water vapour in the air on a damp day. And over busy cities and industrial areas there is more carbon dioxide, as well as poisonous gases such as carbon monoxide and sulphur dioxide.

Where did the atmosphere come from?

The Earth was formed about 4600 million years ago, when gravity caused a hot, dense band of gas and dust around the Sun to collapse in on itself. It got hotter and hotter as particles got pulled in and squashed together. Then it started to cool, solidify, and break up into chunks. The result was the Earth and other planets.

The Earth from space.

The atmosphere is a very thin layer.

Ionosphere
80 km — 85 °C
space shuttle orbits at about 300 km

70 km

Mesosphere

60 km — 1 mb

50 km — 0 °C
research balloons

Stratosphere
40 km — −25° C   10 mb

highest jet aircraft

30 km

natural ozone layer −45° C

supersonic jets
20 km — −70° C   100 mb

highest cloud
jet airliners
10 km

Troposphere
Mount Everest
15° C   1000 mb
sea-level

The four layers in the Earth's atmosphere.

206

Inside, the Earth was very hot and full of activity. Hot gases burst out of it through volcanoes. Slowly, over millions of years, our atmosphere developed from these gases.

4500 million years ago →→→→→ 3500 million years ago

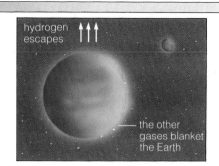

As well as molten rock, volcanoes spewed out water vapour, carbon dioxide, nitrogen, hydrogen chloride, hydrogen . . .

. . . a little argon and tiny amounts of other noble gases. Hydrogen is so light that it escaped to outer space, leaving the rest behind.

The water vapour cooled and condensed and formed the oceans. This is where primitive life first started, about 3500 million years ago.

All the hydrogen chloride and much of the carbon dioxide dissolved in rain and ocean water. This acidic solution attacked rock and wore it away.

Then around 2200 million years ago, the first green plants developed and photosynthesis began. It used up more carbon dioxide, and produced oxygen.

Some of the oxygen reacted with elements on the Earth's surface. The rest went into the atmosphere, to complete the mixture called air.

The ozone layer

The ozone layer is about 25 kilometres above sea-level, in the stratosphere. Ozone has the formula O_3. It is produced when energy from ultraviolet light causes oxygen molecules to break into atoms:

$$O_2 \xrightarrow[\text{light}]{\text{ultraviolet}} 2O \quad \text{then} \quad 2O_2 + 2O \longrightarrow 2O_3$$

The ozone layer protects us from the Sun's most harmful ultraviolet rays.

Questions

1 Make a table to describe the layers of the atmosphere. You should use these headings:
   name   height   temperature   pressure
2 What gas was missing in the early atmosphere?

3 Explain why the atmosphere now contains:
   a more oxygen than 3000 million years ago;
   b less carbon dioxide than 3000 million years ago.
4 Explain how ozone is formed.

13.4 The circulating atmosphere

How the atmosphere is maintained

All living things, including humans, depend on the nitrogen, oxygen and carbon dioxide in air. We also depend on the water vapour that condenses and falls as rain. But these substances don't just get used up and disappear. They are **recycled**.

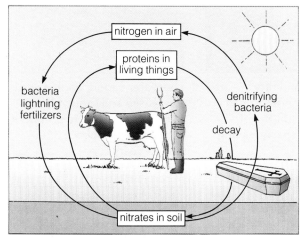

1 Nitrogen circulates between the air, the soil and living things in the **nitrogen cycle**. You'll learn more about this in biology class.

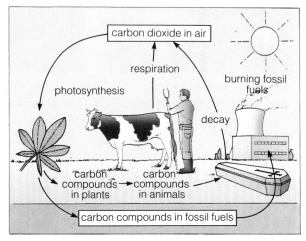

2 Carbon dioxide circulates between the air, the soil and living things in the **carbon cycle**. See page 187 for more about the carbon cycle.

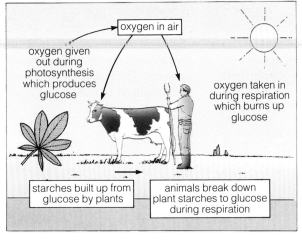

3 The process of photosynthesis, followed by respiration, also recycles **oxygen**. See page 186.

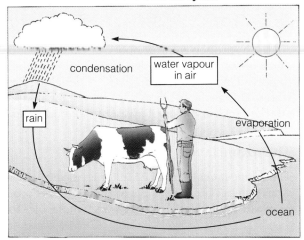

4 And finally, water circulates between the air, the oceans and living things in the **water cycle**.

How the atmosphere protects us

As well as storing the gases living things need, the atmosphere helps us in another very important way: it regulates heat. It acts as a blanket round the Earth, keeping it within temperatures that allow life to survive. It does this by trapping some of the Sun's heat and keeping out the rest. The ozone layer helps to keep out harmful radiation.

With no atmosphere the Moon's temperature varies enormously from day to night.

To help stop the destruction of the ozone layer, we can use products which do not contain CFCs.

How we misuse the atmosphere

For millions of years, nature has maintained the atmosphere by means of the cycles shown opposite. But for the last 150 years, humans have misused it by turning it into a dump for waste gases.

Gases from burning fossil fuels The main waste gas is carbon dioxide. We pour about 5000 million tonnes of it into the air *every year* by burning fossil fuels: coal, oil, diesel and petrol. The oceans can dissolve about 30% of this carbon dioxide. But the rest stays in the atmosphere, upsetting the balance. It traps the Sun's heat and many scientists accept this will make the Earth warmer (page 190). Global warming could mean that the ice in the Arctic and Antarctic will melt, causing huge floods.

Burning fossil fuels also produces many other harmful gases, such as sulphur dioxide and nitrogen oxides, which cause acid rain, and carbon monoxide, which is poisonous. Vehicle engines are the main culprit. More than half the pollution in the air over Britain comes from vehicle exhausts. Find out more about this on page 190.

Gases that destroy the ozone layer Some waste gases destroy the ozone layer. The main ones are chlorine compounds such as chlorofluorocarbons (CFCs). These are used in air conditioning and refrigeration, in aerosols, and as solvents in the electronics industry. The chlorine from them reacts with ozone (O_3), breaking it down to oxygen (O_2). This causes holes in the ozone layer. So more of the Sun's harmful radiation can reach Earth, causing skin cancer and eye cataracts; also crop damage.

Questions

1 Use drawings to help you explain how you personally take part in the cycle that maintains:
a carbon dioxide **b** oxygen
in the atmosphere.

2 What change happened around 1850, which caused us to burn more fossil fuel than we did before?

3 Scientists are working hard to develop compounds that can be used instead of CFCs. Explain why.

13.5 Making use of air

Separating gases from the air

There's a lot of air around. Just as well, since we need it for breathing! But we've got other uses for it too. For example, oxygen from the air is used in making steel, and nitrogen is turned into fertilizers. But first these gases must be **separated** from the air. Every year in the UK alone, over 20 million tonnes of air is separated into its different gases. The process depends on the fact that these gases have different boiling points. This is what happens:

- first the air is cooled down until it turns into a liquid;
- then the liquid air is heated up again. The different gases boil off at different temperatures and are collected one by one. This is an example of **fractional distillation**.

A diagram of an air separation plant is shown below.

Boiling points of gases in air/°C	
Carbon dioxide (sublimes)	−32
Xenon	−108
Krypton	−153
Oxygen	−183
Argon	−186
Nitrogen	−196
Neon	−246
Helium	−269

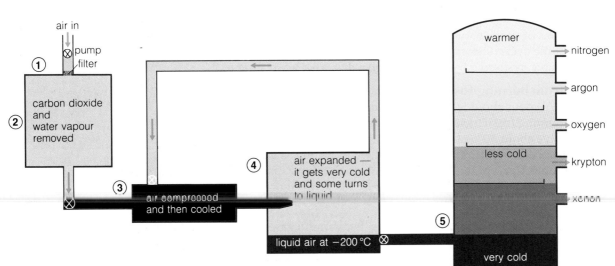

The steps in the process are:

1 Air from outside is pumped into the plant. The filter gets rid of any dust.
2 Next the carbon dioxide and water vapour are removed. Otherwise they would freeze later, when the air is cooled, and block the pipes.
3 Next the air is forced into a small space, or **compressed**. That makes it hot (just like air gets hot when you pump it into a bicycle tyre). It is then cooled down again.
4 The cold, compressed air is passed through a jet, into a larger space. It expands, and that makes it very cold.

Steps 3 and 4 are repeated several times, and each time the air gets colder. By the time it reaches −200 °C, all its gases have become liquid, except for neon and helium. These two are removed.

5 The liquid air is pumped into the fractionating column. There it is slowly warmed up. The gases boil off one by one, and are collected in tanks or cylinders. Nitrogen boils off first. Can you explain why?

Liquid nitrogen.

Making use of oxygen

1 Astronauts have to carry oxygen with them, and so do deep-sea divers. Aeroplanes also carry their own oxygen supply.
2 In hospitals, it is given to patients with breathing difficulties.
3 In steelworks, it is blown through molten steel to purify it (page 159).
4 A mixture of oxygen and ethyne (acetylene) is used as the fuel in oxyacetylene welding torches, because it gives a very hot flame.

Making use of nitrogen

1 It is used for making ammonia, nitric acid and fertilizers.
2 Liquid nitrogen is very cold (− 196 °C) so is used for quick-freezing food in food factories, and freezing liquid in damaged pipes.
3 Because the gas is also unreactive or **inert**, it is pumped into oil storage tanks to reduce the risk of fire, and flushed through food packaging to keep the food fresh.

Making use of carbon dioxide

It is used in freezers, fire extinguishers and to make drinks fizzy. See page 179.

Making use of the noble gases

1 For lighting. **Argon** is used in ordinary lights. **Neon** is used in advertising signs because it glows red when an electric current is passed through it. **Krypton** and **xenon** are used in powerful lamps, such as lighthouse lamps.
2 Some welding uses electricity, rather than a hot flame, to melt the metals. It is often carried out in an atmosphere of argon, to stop the oxygen in air from reacting with the metals.
3 **Helium** is used to fill balloons, since it is both light and unreactive.

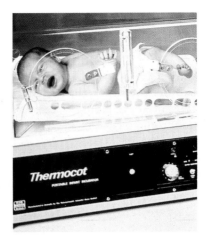
This cot has its own oxygen supply for sick infants.

Frozen meat coming out of a liquid nitrogen freezing tunnel.

Liquid nitrogen is used to freeze liquid in damaged pipes, before repairs.

Questions

1 In the separation of air into its gases:
   a why is the air compressed and then expanded?
   b why is argon obtained *before* oxygen?
2 Write down three uses of oxygen gas.
3 Why does a mixture of oxygen and ethyne burn better than a mixture of air and ethyne?
4 What *two* properties of nitrogen make it suitable for keeping food frozen during transportation?

13.6 Water and the water cycle

What is water?

Water is a compound of hydrogen and oxygen. It has the formula H_2O. As you saw in the last Unit, it first appeared on Earth thousands of millions of years ago as water vapour, in the gases that burst from volcanoes. The water vapour cooled, condensed and formed the oceans.

There's over 361 cubic km of water on Earth, and it covers over 71% of the Earth's surface. About 97% of it is in the oceans. The Pacific Ocean alone covers almost half the Earth. But that's not all. Living things are mostly water. You are over 70% water. Lettuce is about 96% water!

The beginning and end of the water cycle?

More about the oceans

The oceans are giant solutions. They contain some of every element found in the Earth's crust. That is not surprising, because for millions of years rain, streams, rivers and the ocean itself have been wearing the Earth's crust away.

Wind also carries dust and particles far out into the ocean. Gases from the air dissolve in it. Hot fluids rich in dissolved substances erupt from the ocean floor.

Main substances dissolved in ocean water	Mass dissolved in g per kg of water
Chloride ions (Cl⁻)	19.2
Sodium ions (Na⁺)	10.7
Magnesium ions (Mg²⁺)	1.3
Sulphate ions (SO₄²⁻)	2.7
Potassium ions (K⁺)	0.4
Carbon (as carbonate and bicarbonate ions)	0.03

A sink for carbon dioxide? As shown in the table on page 19, carbon dioxide is quite soluble in water. There is much more carbon dioxide dissolved in the oceans than there is in the air.

We now produce huge amounts of carbon dioxide by burning fossil fuels. If the oceans could dissolve it all there might not be a problem. But they can't. The amount is limited by the concentration of carbonate ions at the ocean surface, where this reaction takes place:

$$CO_2 + CO_3^{2-} + H_2O \longrightarrow 2HCO_3^{-}$$

Water and living things

Life began in water: 3500 million years ago the first simple algae and bacteria developed in the oceans. From that time on, water has been essential to life. Every cell in every living thing contains a little water.

Your body contains about 35 litres of water. Most of it is in your cells, the rest in your blood, saliva and other body fluids. Every day, you give out about 2.5 litres of water through sweat and urine, and as water vapour. If you didn't replace this water by eating and drinking, you'd die in a matter of days.

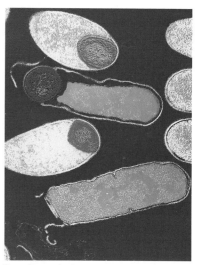

Even bacteria depend on water for survival.

The water cycle

Water circulates between the air, the oceans and living things through the **water cycle**, which is driven by the Sun:

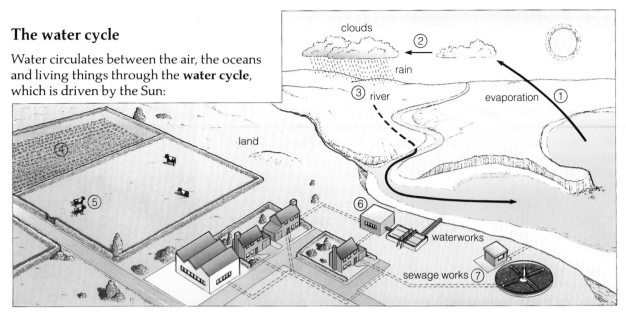

1 Heat from the Sun causes water to evaporate from seas and oceans. The vapour rises, cools and condenses to form tiny water droplets. The droplets form clouds.
2 The clouds get carried along on air currents. They cool, and the droplets join to form larger drops which fall as rain. Or, if the air is very cold, as hail, sleet or snow.
3 Some rainwater soaks through the ground, and reappears as springs. Some flows along the ground as streams. The springs and streams feed rivers. The rivers flow to the sea. The main cycle is complete.

So where do living things come into the picture?

4 Plants soak up rainwater through their roots, and use it for photosynthesis and building their cells and fluids. They give out water vapour during respiration.
5 Animals take in water by eating and drinking. They give it out in sweat and urine, and breathe out water vapour from respiration.
6 River water is filtered, cleaned up and pumped into homes and factories for washing and cooking, or use as a raw material.
7 Waste water from homes and factories is filtered and cleaned up at sewage works, and pumped back into the river.

Stages 6 and 7 are where problems arise. We have dumped all kinds of waste into our water over the years: waste chemicals, raw sewage, fertilizer, organic solvents, and detergents. These can't all be removed at waterworks or sewage works. They harm fish and other river life. They also harm humans. For example, scientists think that certain chemicals in drinking water may be lowering the sperm count in males, which means less chance of having children.

The 'cycle' is not at work here.

Questions

1 Why is there so much chloride ion in the oceans? (Hint: look again at page 18.)
2 Explain why the oceans can't dissolve all the carbon dioxide produced by burning fossil fuels.
3 Ocean water is salty. Use the water cycle to help you explain why river water is *not* salty.
4 Write a list of the ways you personally get involved in the water cycle.

213

13.7 Our water supply

Where tap water comes from

In Britain, some tap water comes from **rivers**, some from **underground wells**, and some from **mountain reservoirs**. Water from these sources is never completely pure, especially river water. It may contain:

1 **Bacteria** Tiny living organisms, so small you would need a microscope to see them. Most bacteria are harmless, but some can cause disease.
2 **Dissolved substances** Nitrates and sulphates from the soil, gases from the air, and sometimes calcium and magnesium compounds from rocks.
3 **Solid substances** Particles of mud, sand, grit, twigs, dead plants, and perhaps tins and rags that people have dumped.

Before the water is safe to drink, the bacteria and solid substances must be removed. This is done at the **waterworks**.

Part of a modern waterworks.

The waterworks

This diagram shows what happens at a waterworks:

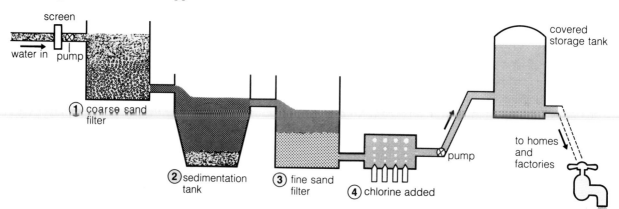

The water is pumped in through a screen, which gets rid of the larger bits of rubbish. Then it goes through these stages:

1 It is filtered through a bed of coarse sand, which traps the larger particles of solid.
2 Next it flows into a **sedimentation tank**. Here chemicals are added to it, to make the smaller particles stick together, These particles then settle to the bottom of the tank.
3 Water flows from the top of the sedimentation tank, into a filter of fine sand. This traps any remaining particles.
4 Finally a little chlorine gas is added. It dissolves, and kills any remaining bacteria. This is called **disinfecting** or **sterilizing** the water.

In some places, a fluoride compound is also added to the water, to help prevent tooth decay. The water is now fit to drink. It is pumped into high storage tanks, and from there piped to homes and factories.

Fluoride added to the water supply helps to ensure healthy teeth.

Waste water and sewage plants

All sorts of things get mixed with tap water: shampoo, toothpaste, detergents, grease, body waste, food, grit, sand, and waste from factories. This mixture goes down the drain, and is called **sewage**. It is piped underground to a **sewage plant**, where the water in it is cleaned up and fed back to the river. Below is a diagram of the plant:

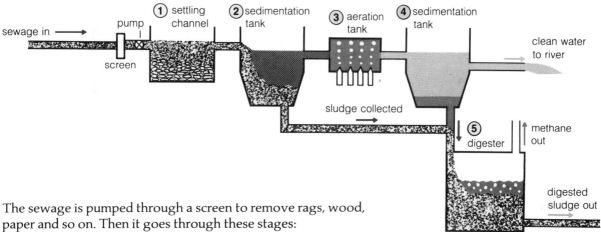

The sewage is pumped through a screen to remove rags, wood, paper and so on. Then it goes through these stages:

1 It flows slowly along a **settling channel**, where grit and sand settle out.
2 Next it passes into a **sedimentation tank**. Here smaller pieces of waste sink slowly to the bottom. This waste is called **sludge**. It is grey and evil-smelling, and contains many harmful things.
3 The water now looks cleaner. It flows into an **aeration tank**, which contains special bacteria growing on sludge. These bacteria feed on harmful things in the water, and make them harmless. For this they need a lot of oxygen, so air is continually pumped through the sludge, from the bottom of the tank. Instead of aeration tanks, some plants use **percolating filters**, where the bacteria live on stones, and the water trickles over them.
4 Next comes another **sedimentation tank**, where any remaining sludge settles out. The water is now safe to put into the river.
5 All the sludge is collected into tanks called **digesters**. Here it is mixed with bacteria which destroy the harmful substances, producing **methane gas**. Methane is a good fuel, so may be used to make electricity for the sewage plant. The digested sludge is burned to ash or sold to farmers as fertilizer.

As you can see, the steps above remove solid waste and many other harmful substances. But they can't remove everything. Some harmful substances in the waste from homes and factories will end up in the river, and perhaps even back in tap water. Not good for fish or humans!

Bacteria are at work in these filter beds.

Questions

1 List ten impurities that you might find in river water.
2 What happens in the sedimentation tanks at waterworks?
3 What is: **a** sewage? **b** sludge?
4 At a sewage plant, describe what happens in:
 a aeration tanks **b** digesters

13.8 Soft and hard water

As you saw on page 214, water is treated at the waterworks, before being piped to your home.

The treatment removes only the insoluble particles and kills bacteria. So the water still isn't pure. Depending on its source, it may contain chemicals from waste dumped from homes, farms and factories. It also contains compounds dissolved from rocks and soil: mainly calcium sulphate, calcium hydrogen carbonate, magnesium sulphate, and magnesium hydrogen carbonate.

How much of these calcium and magnesium compounds are in your tapwater? It all depends on where you live. Soap will give you a clue:

Washday blues. Soap causes a scum in hard water which is difficult to wash out of clothes. Modern detergents do not contain soap, so do not make scum.

lather

soap flakes added

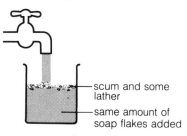

scum and some lather

same amount of soap flakes added

If soap lathers easily, it means the water contains very little calcium and magnesium compounds. It is **soft** water.

But a greyish scum and hardly any lather shows that larger amounts are present. The water is called **hard** water.

The scum forms because the calcium and magnesium compounds react with soap to give an insoluble product. For example:

calcium sulphate + sodium stearate $\longrightarrow$ calcium stearate + sodium sulphate
                (soap)   (scum)

Where hardness comes from

Calcium hydrogen carbonate is the main cause of hard water. It forms when rain falls on rocks of **limestone** and **chalk**. These are made of calcium carbonate which is *not* soluble in water. But rainwater contains carbon dioxide dissolved from the air, which makes it acidic. So it reacts with the rocks to form calcium hydrogen carbonate which *is* soluble and ends up in our taps. The reaction is:

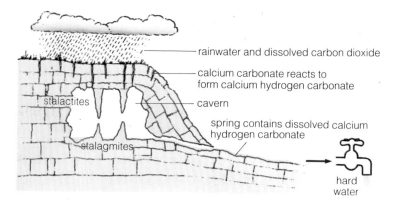

rainwater and dissolved carbon dioxide

calcium carbonate reacts to form calcium hydrogen carbonate

stalactites

cavern

spring contains dissolved calcium hydrogen carbonate

stalagmites

hard water

water + carbon dioxide + calcium carbonate $\longrightarrow$ calcium hydrogen carbonate
$H_2O\,(l)\ +\ \ \ \ CO_2\,(g)\ \ \ +\ \ \ \ CaCO_3\,(s)\ \ \ \longrightarrow\ \ \ \ \ \ Ca(HCO_3)_2\,(aq)$

This reaction between rainwater and rocks is part of the **weathering** process. You can find out more about it on page 218.

In the same way, rainwater reacts with other rocks such as **dolomite** ($CaCO_3.MgCO_3$) and **gypsum** ($CaSO_4.2H_2O$) to give the other compounds that make tapwater water hard.

Stalactites and stalagmites

In the diagram on the opposite page, there are **stalactites** and **stalagmites** in the underground cavern. These are often found in limestone areas.

A stalactite hangs from the cavern roof. (It hangs on *tight*!) It starts to form when drops of water containing calcium hydrogen carbonate collect on the roof. The water evaporates, losing carbon dioxide, and solid calcium carbonate is left behind:

$$\text{calcium hydrogen carbonate} \longrightarrow \text{water} + \text{carbon dioxide} + \text{calcium carbonate}$$

$$Ca(HCO_3)_2\,(aq) \longrightarrow H_2O\,(g) + CO_2\,(g) + CaCO_3\,(s)$$

This carries on, drop by drop, on the same spot. Slowly, over thousands of years, a spike of calcium carbonate grows down. A stalagmite forms in exactly the same way, but grows in the opposite direction.

Stalactites and stalagmites.

From rocks to kettles

 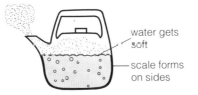

kettle of hard water

water gets soft

scale forms on sides

This water contains calcium hydrogen carbonate: it is hard. When the water is boiled, the compound breaks down to form calcium carbonate which is . . .

. . . insoluble. It forms a hard **scale** inside the kettle. But the water itself is now soft since the calcium hydrogen carbonate has been removed.

The same reaction occurs inside boilers and pipes in factories and central heating systems. Scale has blocked this pipe almost completely.

Since it can be removed just by boiling, the hardness caused by calcium hydrogen carbonate is called **temporary hardness**. The reaction is exactly the same as the one that forms stalactites. Hardness caused by other compounds is called **permanent hardness** because boiling does not affect it. But it can be removed in other ways, for example by adding sodium carbonate (washing soda).

Questions

1 Hard water wastes soap. Explain why. Use an equation to help you.
2 Chalk is insoluble in water. Why does it dissolve in rainwater? Write an equation for the reaction.
3 Explain how a *stalagmite* forms.

4 When washing soda is added to hard water containing calcium sulphate, calcium carbonate is precipitated. Write an equation for the reaction.
5 Temporary hardness costs money. Write down as many reasons as you can to explain why.

13.9 Weathering

The rocks at the surface of our planet are under attack. Too slowly for us to notice in everyday life, rocks are broken down into smaller bits or dissolved in the water around them. The process is called **weathering**.

Types of weathering

Weathering processes cause changes to the rocks. These changes are either physical or chemical.

Physical weathering   This happens when forces open up any weak points in the rock. For example:

1 When water turns to ice it expands. If this happens in cracks in rocks, the cracks widen.
2 Rocks are mixtures of different minerals. When they are heated by the Sun, these minerals expand by different amounts. Each grain pushes against the next, making the rock weaker.
3 When a seed falls into a crack in a rock, it may find conditions good enough to allow germination. A stem grows upwards and a root grows down. The growing plant forces the crack to widen.

Chemical weathering: rocks attacked by the weak carbonic acid in rainwater.

Chemical weathering: rusty rocks.

Physical weathering: fragments broken off by expanding ice.

Chemical weathering This happens when there is a chemical reaction between the rock and its surroundings. For example:

1 Rainwater picks up carbon dioxide from the air, so it is slightly acidic . The acid attacks some rocks, especially limestone.
2 Ground water becomes acidic if it runs through soil containing rotting plants. These make acid which also attacks limestone.
3 Some rocks contain iron compounds. The oxygen and water in the environment can react with the iron to make rust.

Weathered landscapes

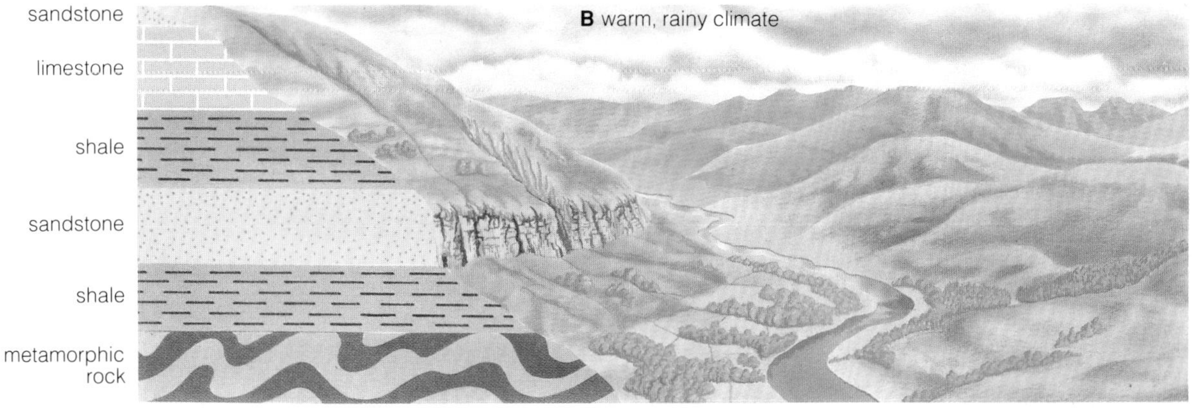

sandstone
limestone
shale
sandstone
shale
metamorphic rock

A hot, dry climate

sandstone
limestone
shale
sandstone
shale
metamorphic rock

B warm, rainy climate

The landscapes above have exactly the same layers of rock. In the hot, dry climate, physical weathering has had the most effect. Steep cliffs have formed with loose fragments at their base. In the warm, rainy climate, chemical weathering has had the most effect. Chemical solutions have been able to form because of the water.

Questions

1 What is meant by *physical weathering*? Give two examples of physical weathering.
2 What is meant by *chemical weathering*? Give two examples of chemical weathering.
3 Which type of weathering would you most expect to find where the climate is warm and wet? Explain your answer.
4 Why can frost cause weathering?

13.10 Soil and sediment transport

Soil formation

Physical weathering produces rock fragments of all sizes. If these fragments stay in place, they form the starting material for soil. From bare rock, soil takes about 1000 years to form by the action of weathering processes and the effects of plants and animals.

A vertical section through soil is called a **soil profile**. The top layer, the **topsoil**, has **humus** mixed in with its soil particles. Humus is the decayed remains of plants and animals, and it contains the substances which plants need to grow. It also gives soil its dark colour. Underneath the topsoil is the **subsoil**. This has no humus, so its colour is lighter. It has more of the larger rock fragments which we call stones.

A soil profile.

Transport of weathering products

If there is enough force acting on the products of weathering, then they will move.

Gravity is one force which moves rock fragments. If gravity is the only force acting, then rock fragments on a flat surface stay in place. On slopes they may be held in place by friction. But if the slope is steep enough, friction cannot hold the fragments. **Landslips** happen when loose material moves under the effect of gravity.

Gravity may act on its own, but water, wind, and ice are the main **transport agents** for carrying sediment (pebbles, sand, and mud). They remove fragments, causing erosion.

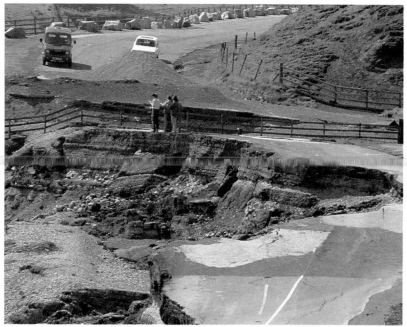
This road has collapsed because the underlying rock was not strong enough to support the load placed on it.

Water transport

Water can carry sediment over the land either in **channels** or in **sheet wash**. Chemical solutions can be transported by channels and sheet wash, and also by ground water.

The graph on the opposite page shows the results of experiments on sediment movement in water. The top curve shows the speeds at which sediment is removed. The bottom curve shows the speeds at which sediment is deposited on the bed of the channel.

Pebbles, sand, and mud only move if they have enough energy. The larger the particle, the more energy is needed to transport it.

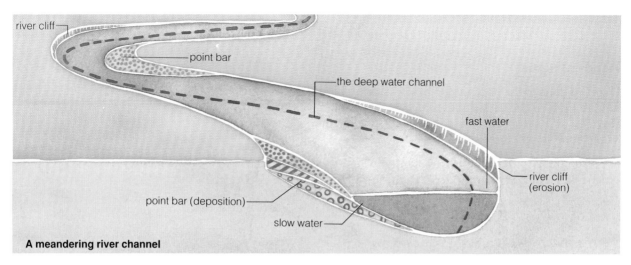

A meandering river channel

Rivers and streams usually develop curved channels, or **meanders**. Their banks are worn away on the outside of each bend, where the flow is fastest. The bigger particles are dropped on the inside of the bend, where the flow is slowest.

The effect may be investigated as shown on the right. As the water is pushed round, the movement of the sand can be studied.

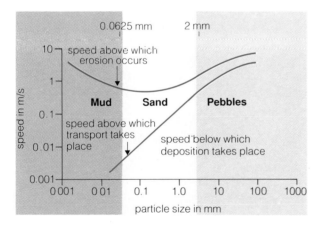

How moving water affects mud, sand, and pebbles.

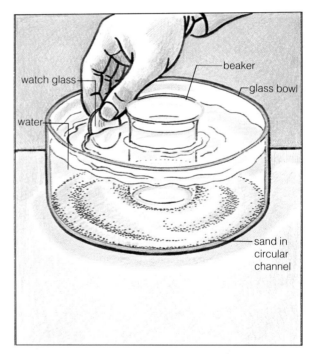

Questions

1 How is soil formed?
2 What is a *soil profile*?
3 Give two examples of transport agents.
4 When a river meanders:
  a On which bank is the flow fastest?
  b Which bank is worn away most?
  c Near which bank is most sand dropped? Explain your answer.

5 Use the graph on this page to answer the following:
  a What is the slowest water speed which can lift 1 mm grains of sand from stream beds?
  b What size of particle could be picked up by a river flowing at 1 m/s?
  c If particles of mud, which are about 0.01 mm big, have settled on a stream bed, what can you say about the speed of the water?

13.11 Transport and rock formation

Wind transport

The sand dune in the photograph below is being blown by the wind. Sand is lifted off the back slope, carried over the top, then dropped down on to the front to make a steep-angled layer. In this way, the whole dune slowly moves across the desert.

The sand grains in the dune are constantly being rubbed together, which makes them much rounder than those in river sand.

Ice transport

Glaciers and ice sheets move very slowly down slopes. As they do so, they collect frost-shattered fragments from the rocks around them and carry them along. The fragments can fall through cracks into the ice. As the ice moves, it wears more fragments out of the rocks below. These add to the 'ice file' rubbing away at the earth. When the ice melts, the fragments in the ice begin to wash out.

Sand dunes move across the desert as grains of sand are carried by the wind from one side to the other.

Rock fragments carried by this glacier are deposited when it starts to melt.

From sediments to rocks

Rock fragments and chemical solutions may turn into sedimentary rocks when deposited – though the process takes millions of years. It can happen in deserts, seas, or rivers, or other **depositional environments**. Chemicals soaking through the sediment help turn them into new rocks, because they cement the grains together.

Three different types of sediment (pebbles, sand, and mud) and the different types of sedimentary rock which are formed from them.

Rock fragments may move around for some time before being deposited, or they may settle quickly. Looking at how fragments are deposited can give clues about how different sedimentary rocks were formed.

The bar graphs above show the results of sieving beach sand and glacial sand. The glacial sand is **badly sorted** – it has a wide range of sediment sizes. The beach sand is **well sorted**.

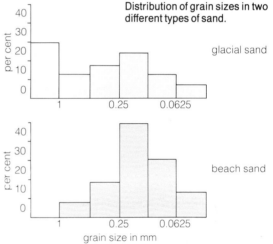

Distribution of grain sizes in two different types of sand.

glacial sand

beach sand

grain size in mm

Some sedimentary rock types

Sedimentary rocks made of fragments are given names based on the average size of the fragments in them.

Rocks made of mud particles, less than 0.0625 mm across (1/16 mm), are called **shales** if they have thin layers, or **mudstones** if they do not have layers. **Sandstones** are made of particles with sizes from 0.0625 mm up to 2 mm. Rocks made of pebbles over 2 mm in size are called **conglomerates**. Some very badly sorted sedimentary rocks can have a range of grain sizes from pebbles to mud. These are often called **greywacke sandstones**.

Questions

1 Why are grains of dune sand rounder than those of river sand?

2 Rock fragments dropped by ice have sharp corners and are often scratched. Suggest reasons why.

3 What do the bar graphs above tell you about the difference between the grains in glacial sand and those in beach sand?

4 *shale   mudstone   sandstone   conglomerate*
Which of the sedimentary rocks above:
a Contains large pebbles?
b Is formed from mud, and is laid down in thin layers?
c Contains particles with sizes between 0.0625 mm and 2 mm?

13.12 Sedimentary environments

Depositional environments

In some cases, the edges of continents are underwater. The underwater part is called the **continental shelf**. Beyond it, the sea-bed falls away steeply into the depths of the ocean.

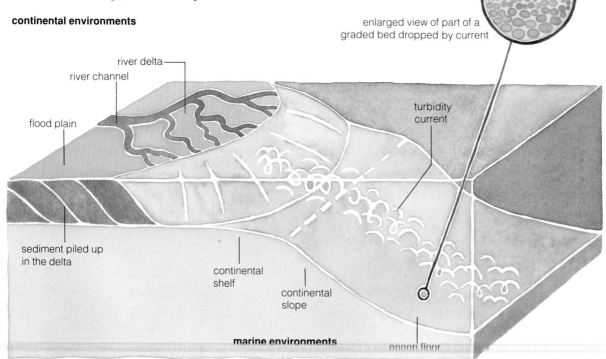

continental environments

river delta

river channel

flood plain

sediment piled up in the delta

continental shelf

continental slope

turbidity current

enlarged view of part of a graded bed dropped by current

marine environments   ocean floor

How some sediments reach the ocean floor.

Continental environments   On the continents, rivers may deposit sediment in channels and on flood plains.

Marine environments   Sediment may be carried out to the continental shelf. When a river reaches the sea, the water slows and sediment is dropped. It makes flat layers called **beds**. The continental shelf and the deep sea-bed are both marine environments.

Intertidal environments   If sediment builds up above the water level of the sea, it can form deltas. Delta tops and beaches are above water at low tide, but below water at high tide. They are intertidal.

Depositing sediment

As sediment piles up on the continental shelf, it can slip off to form an underwater avalanche. This mixture of water and sediment flows down the slope, getting faster as it does so, before rushing across the ocean floor as a **turbidity current**. As the flow slows down again, the larger particles drop first. The light mud particles cannot drop until the flow stops. They make layers of **graded bedding**, with bigger particles at the bottom.

Ripples on a beach.

224

Ripples form when water moves sediment. **Dunes** are large ripples made in faster, deeper flows. They form both underwater and as wind deposits (see page 222). A cross-section of a dune shows layers tilted at angles. This is called **cross bedding**.

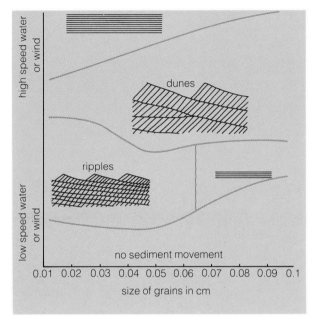

How ripples and dunes form.

Part of a coral reef.

Chemical and organic sediments

Evaporites   These are made when chemical solutions dry up, leaving salts behind in layers of crystals.

Sea water is a solution containing many salts. Although we consider calcium carbonate ($CaCO_3$) insoluble, very tiny amounts do dissolve. It will form crystals in warm, shallow water, as the water evaporates.

Corals   Some living things use chemicals from sea water to make hard parts such as shells or bones. Corals are tiny organisms which build up large underwater structures, called **reefs**, from calcium carbonate. To live and grow, corals require warm, shallow sea water, with plenty of light.

Chalk   This is a type of limestone, formed by single-celled organisms called **algae**. They extract the calcium carbonate from sea water to make tiny armour plates. These plates collect on the sea-bed when they die.

Salts in sea water		
Chemical	Percentage	Order of solubility
Mg and K salts	18.1	1
NaCl	78.0	2
$CaSO_4$	3.5	3
$CaCO_3$	0.3	4

Questions

1 Why might sediment start moving off a continental shelf and into deeper water?
2 Use the chart on this page to decide what is the biggest grain size that will form ripples.
3 In what kind of climate would evaporites form?
4 Look at the table above. If some water was evaporated away completely, in what order would the chemicals form crystals on the sea-bed?
5 What is the main chemical substance in:
   a a coral reef?   b chalk?

13.13 Interpreting sedimentary layers

Sedimentary rocks cover 75 per cent of the surface of the continents. Their layers contain the evidence for events that shaped the landscape over millions of years.

Evidence from layers

The photograph below shows some limestone. It is part of a layer 30 metres thick. This contains oval lumps, several metres across, made up of large corals and other fossils. In between the lumps are flat **beds** (layers) of limestone, mainly made of fossil shells and broken fossils. The presence of the coral suggests that the rock was formed in a shallow sea. The types of fossil indicate that the rock is of Silurian age, over 400 million years old.

Limestone with fossils.

Sandstone in a quarry.

The photograph above right shows some sandstone in a quarry. If you could look closely at the grains in the rock, you would find that they were round and well sorted (of similar size). This suggests that the rock formed from a sand dune in a desert. The curved, sloping surfaces of the beds also support this idea. Various evidence dates the rock as about 270 million years old.

The diagram on the right shows part of the face of a hill in which layers of sandstone alternate with layers of shale. The sandstone layers have the largest grains near the base, and ripples on the top. The shales have grooves in their top surface where the next sandstone begins.

The nature of the layers suggests that the rocks were formed in deep sea from material carried by moving water. Particles for the sandstone were deposited by turbidity currents, and have made layers of graded bedding.

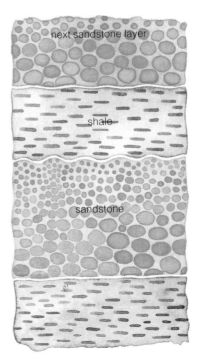

next sandstone layer

shale

sandstone

Coal measures

Studying sedimentary rocks on their own can tell us a great deal, but looking at a piled-up sequence of sedimentary rocks can often tell us more.

Coalfield areas provide good examples of patterns of layers. These rocks are mostly mudstones, with some sandstones and, of course, coal seams as well. The coal seams are only a small part of these sets of rocks, which are called **coal measures**. They are about 300 million years old.

Coal measures can be divided into **cycles**, as shown in the diagram on the right. Each cycle ends with a coal seam.

Here are a number of extra observations about the rocks in coal measures.

1 The black shale layer at the start of each cycle, just above the last coal seam, sometimes contains fossil sea shells.
2 Sandy shale contains a mixture of sand and mud.
3 The sandstones sometimes have cross bedding.
4 The layer under the coal seam in each cycle often contains fossil roots. This layer is often very pale in colour.
5 The coal seams contain fossils of tree bark and branches.

The diagram below gives more information about one coal measure cycle. The column headed *Interpretation* shows how scientists think the layers in each part of the cycle were formed.

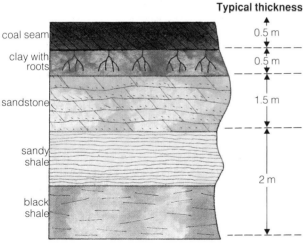

	Typical thickness	Interpretation
coal seam	0.5 m	plants in a swamp forest
clay with roots	0.5 m	flooded soil below the plants, with freshwater shellfish
sandstone	1.5 m	rivers washed sand across the shallow swamps, making them very shallow
sandy shale	2 m	
black shale		shallow sea which flooded across the last swamp forest

Key

- coal
- white clay with fossil roots
- sandstone with cross bedding
- sandy shale
- black shale

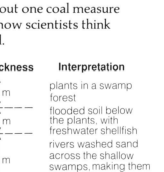

one cycle

Questions

To answer the following, you may need information from earlier Units.

1 What evidence is there that the limestone described on the opposite page formed in shallow sea?
2 What evidence is there that the sandstone in the photograph on the opposite page formed from a desert dune?

3 In the diagram on the opposite page:
  a What do the different grain sizes tell you about how the sandstone was formed?
  b What do the ripples on the sandstone tell you about how it was formed?
4 Why does the layer below a coal seam often have fossil roots in it?

Questions on Section 13

1 Give reasons for each of the following:
a Scientists have divided the Earth's past into separate geological time periods.
b In sedimentary rocks, the deepest layer is likely to be the oldest.
c Scientists think that the Earth must be at least 4000 million years old.

2 Copy and complete the following paragraph:
Air is a . . . . of different gases. 99% of it consists of the two elements . . . . and . . . . One of these, ..., is needed for respiration, which is the process by which living things obtain the . . . . they need. The two elements above can be . . . . from liquid air by . . . ., because they have different . . . . Much of the . . . . obtained is used to make nitric acid and fertilizers. Some of the remaining 1% of air consists of two compounds, ... and . . . . One of these is important because it is taken in by plants, in the presence of . . . ., to form . . . . The rest of the air is made up of elements called the . . . . These are all members of Group . . . . of the periodic table.

3 The mixture of gases we call air developed over millions of years from the gas that burst from volcanoes.
a Most of the volcanic gas consisted of water vapour. But now there is only a tiny percentage of this in air. Why?
b The volcanic gas contained a large percentage of carbon dioxide. Where did most of this go?
c The volcanic gas also contained hydrogen chloride. Why is there none of this gas in air?
d Hydrogen is also absent from air, even though it was a constituent of volcanic gas. Where has it gone to?
e Which important gas in air was not present in the volcanic gas? Name the process that produced this gas over millions of years.
f Which gas makes up approximately 78% of the air?
g Only one gas in the mixture will allow things to burn in it. Which gas is this?
h How are the gases in the mixture separated from each other, in industry?
i Which noble gas is present in the greatest amount in air?
j Which gas containing sulphur is a major cause of air pollution?
k Name two other gases which contribute to air pollution.
l Name one substance which is not a gas but which also pollutes the air.

4 Oxygen and nitrogen, the two main gases in air, are both slightly soluble in water. Using the apparatus below, a sample of water was boiled until 100 cm³ of dissolved air had been collected.

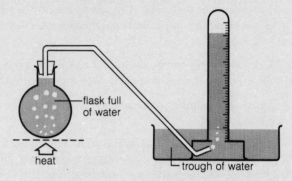

flask full of water
heat
trough of water

This air was then passed over heated copper. Its final volume was 67 cm³.
a Is air *more soluble* or *less soluble* in water, as the temperature rises? How can you tell?
b The copper reacts with the oxygen in the dissolved air. Write an equation for the reaction.
c What volume of oxygen was present in 100 cm³ of dissolved air?
d Calculate the approximate percentages of oxygen and nitrogen in dissolved air.
e What are the percentages of these gases in atmospheric air?
f Explain why your answers are different, for questions d and e.
g Which gas is more soluble in water, *nitrogen* or *oxygen*?
h What is the biological importance of air dissolved in water?

5 Answer the following questions about water.
a Tap water does *not* boil at exactly 100 °C or freeze at exactly 0 °C.
   i Explain why.
   ii Would you expect its freezing point to be higher or lower than 0 °C?
   iii Would you expect its boiling point to be higher or lower than 100 °C?
b Name four compounds that make water hard.
c Name four rocks that cause hardness in water.
d Calcium hydrogen carbonate causes *temporary* hardness in water. Explain what that means.
e Write an equation to show what happens to the calcium hydrogen carbonate when tap water is boiled.
f One way to remove hardness from water is to distill it. Explain how distillation makes hard water pure.

6 Copy and complete:

The carbon cycle and the water cycle are both driven by . . . . from the . . . . In the . . . . cycle, energy in the form of . . . . causes ocean water to . . . . When it cools, the vapour forms . . . . which collect as . . . . Eventually this falls as . . . . This soaks into the ground and is taken in by the . . . . of plants, which need it for a process called . . . . This process is the start of the . . . . cycle. It requires energy in the form of . . . .

7 Hard water is found in many parts of the country. It is formed because rainwater is weakly acidic.
a What gas from the atmosphere dissolves in rainwater to make it weakly acidic?
b Why is water hard in some areas and soft in others?
c Write the chemical equation for the formation of hard water.
d Which geological feature is caused by this reaction?
e Describe how you could use a soap solution to tell hard and soft water apart.
f Which of the following compounds, if added to pure water, would make the water hard?
calcium; magnesium chloride;
sodium chloride; potassium chloride.
g What problem is caused when hard water is used in domestic kettles?
h Write the equation for the chemical reaction that takes place in the kettle.
i What geological feature, which takes thousands of years to complete, is caused by the same chemical reaction?

8 a What is meant by weathering?
b Describe three types of physical weathering.
c Describe three types of chemical weathering.

9 a Draw and label a soil profile.
b Explain how soil forms.
c Why is there often no soil on steep mountains?

10 Explain these discoveries made during experiments on the transport of rock fragments:
a Transportation causes the rounding of rock fragments. The rounder the fragment, the further it has been transported.
b Transportation creates an increasing number of rock fragments.
c Transportation reduces the average size of the rock fragments.
d Transportation sorts rock fragments. The more the sediment has been transported, the smaller the range of sizes of particles.

11 How are the following sedimentary structures formed?
a Graded bedding.
b Ripples.
c Dunes.

12 These diagrams show microscopic views of two sandstones:

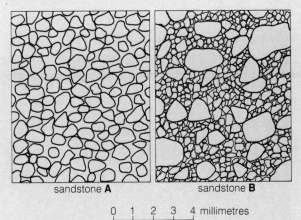

sandstone **A**   sandstone **B**

0 1 2 3 4 millimetres

a Which is the best sorted sandstone?
b Which sandstone has had the most energy acting on it during transport?

13 Imagine that you are visiting a quarry for sedimentary rocks. Describe how you would tell if the rocks were formed in:
a a desert;
b a swamp forest;
c a glacial area.
(Include labelled diagrams with your answers.)

14 Here is a cross-section of some rock layers:

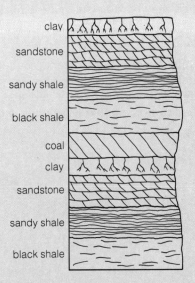

clay
sandstone
sandy shale
black shale
coal
clay
sandstone
sandy shale
black shale

a Which layer is the oldest?
b Which layer formed from the remains of dead plants?
c Which layers were formed from mud particles deposited when a shallow sea flooded a swamp forest?

14.1 Volcanic eruptions

Materials from volcanoes

Agglomerate.

Basalt lava.

The materials in the photographs above came from volcanoes. The **agglomerate** began as rock fragments blown out of the volcano. The **lava** was molten (melted) material that came out of the volcano, cooled down, and crystallized as a solid.

As lava cools its colour changes. The hottest melts are white-hot. As they cool, they become yellow, orange, and then red-hot. As they crystallize, they become dull red and finally black. The most common lava is basalt. When it is erupted, it has a temperature of about 1200 °C. It crystallizes to a solid at about 1150 °C.

High-pressure gas pushes materials out of volcanoes. Different volcanoes give out different mixtures of gases. The main volcanic gases are steam, carbon dioxide, nitrogen, sulphur dioxide, hydrogen, carbon monoxide, sulphur, and chlorine.

Scientists collecting gases from Mount Etna in Sicily.

Part of a solidified lava flow.

Comparing volcanoes

When Mount St. Helens in Alaska erupted in 1980, it had been **dormant** ('sleeping') for over 100 years. It had not erupted during that time. At first, it threw out small amounts of volcanic ash. Then, suddenly, it blasted gas and ash out sideways, turning forests into wasteland.

Mount St. Helens is probably how you expect a volcano to look, but not all volcanoes have a steep cone. Some have a much flatter shape.

Hawaii has volcanoes. Some of them erupt almost constantly, sending out lots of runny lava.

One way of comparing the shapes of volcanoes is to measure the average **slope angle**. The slope angles of Mount St. Helens and a volcano in Hawaii are marked on the photographs on the right.

Volcanoes like Mount St. Helens erupt violently after long, quiet periods. They throw out ash, which piles up around the vent to make a steep cone. They do not erupt very much lava. The lava is sticky because it has a lot of gas in it, and cools quickly.

Other volcanoes, like those on Hawaii, erupt runny lava most of the time. The lava is runny because it has less gas, so heat loss to the air is slower. Lava runs from the vent to make a flattish cone.

Mount St. Helens, Alaska.

A volcano on Hawaii.

Questions

1 What is the most common type of lava?
2 What pushes material out of a volcano? What evidence is there for this in the photograph of the basalt lava on the opposite page?
3 If a volcano is *dormant*, what does this mean?

4 a Use a protractor to measure the slope angles of the two volcanoes in the photographs above.
  b Why does one of the volcanoes have a greater slope angle than the other?
  c How do the two volcanoes differ in behaviour?

14.2 Igneous rocks

Molten rock is called **magma**. When magma cools and solidifies, it forms **igneous** rocks. Sometimes, magma is erupted from volcanoes as lava, which then cools and solidifies. Sometimes, magma cools and solidifies underground.

Extrusive rocks are those which solidify when extruded (pushed out) from the ground.

Intrusive rocks are those which solidify beneath the surface.

Igneous rock: Whin Sill in northern England.

Cooling and crystals

Magma takes longer to cool and solidify underground because of the insulating effects of the surrounding rocks. Near or above the surface, molten rock cools more quickly.

The two igneous rocks in the photographs below came from nearby places on one island. So they probably began as the same magma. They both consist of tiny grains (crystals). However, one of the rocks has a bigger grain size than the other.

This rock is **basalt**. It was erupted on to the surface as lava which then cooled quickly. This meant that there was not time for large crystals to grow. So the grain size is small.

This rock is **gabbro**. It formed from magma which cooled and solidfied slowly underground. This meant that there was time for large crystals to grow. So the grain size is large.

Sills and plutons

The Romans built part of Hadrian's Wall along a ridge in northern England. Features like this ridge are called **sills**. They are layers of igneous rock found sandwiched between layers of sedimentary rock. The igneous rock was squeezed between the layers when it was molten. The sill under Hadrian's Wall is called the Whin Sill. The igneous rock is **dolerite**. It is a dark rock, made of crystals that can only just be seen by the naked eye.

Bodmin Moor in Cornwall is made of igneous rock. It was once covered by sediments which have worn away. Several areas like it are linked underground. Large masses of igneous rock like this are called **plutons**. The pluton under Bodmin Moor is made of **granite**. It has large pale crystals.

Cornish granite.   Dolerite from Whin Sill.

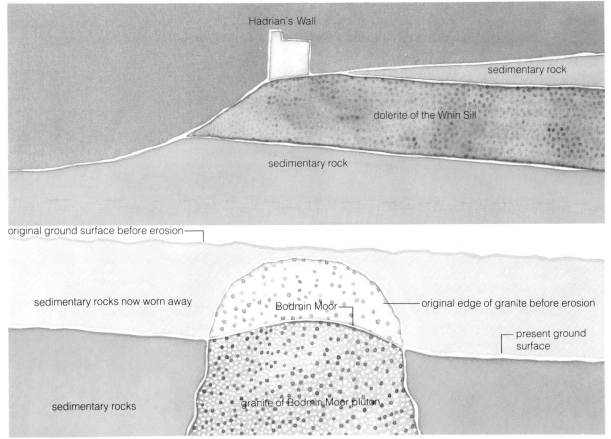

Questions

1 What is the difference between *extrusive* rock and *intrusive* rock? Which type is basalt?
2 What is meant by each of the following?
  a Magma? **b** A sill? **c** A pluton?
3 In the photographs on the opposite page, why does the basalt have a smaller grain size than the gabbro?

4 The photograph at the top of this page shows Cornish granite from Bodmin Moor and dolerite from the Whin Sill. Both formed from molten magma.
  a Which of the rocks do you think cooled more slowly when it formed? How were you able to tell?
  b Why do you think one of these rocks cooled more slowly than the other when it formed?

14.3 Zones of earthquakes and volcanoes

Looking for patterns

We find igneous rocks in most places around the Earth, but *active* volcanoes are not found everywhere.

There are about 500 active volcanoes at present. The map below shows where they are found. Some are under the oceans, and some are on the continents. Some are of the **explosive** type that erupt mostly ash. Others are of the **quiet** type that erupt mostly basalt lava, and keep erupting it most of the time.

The map also shows the location of the main earthquake zones. Note that there are patterns connecting volcanoes and earthquakes. Both tend to be concentrated in the same narrow regions. These include the underwater mountain ranges known as **ocean ridges**, and also the deep **ocean trenches** and **island arcs**.

On ocean ridges, volcanic vents called **black smokers** release dark liquids into the ocean. Scientists have found strange life forms round these vents, including bacteria and huge tubeworms.

Volcanic activity deep in the ocean: a black smoker.

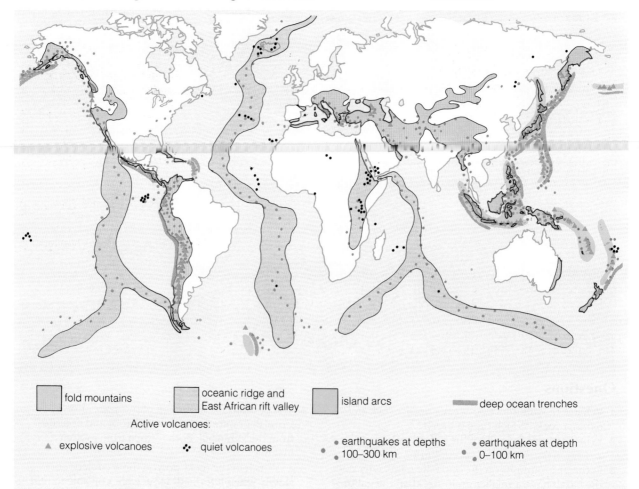

| | fold mountains | | oceanic ridge and East African rift valley | | island arcs | | deep ocean trenches |

Active volcanoes:

▲ explosive volcanoes   ⬩ quiet volcanoes   • earthquakes at depths 100–300 km   • earthquakes at depth 0–100 km

Earthquakes

Earthquakes happen when huge, natural forces affect rocks so that they break and move. A break in a rock is called a **fault**. The breaking rocks release energy into the Earth, which shakes. The place in the ground where the rock breaks is called the **focus** of the earthquake. The point on the surface directly above the focus is the **epicentre**.

Once rock has broken and moved, it can move again if new forces build up around it. Some faults move many times. Earthquake zones are usually places where there are many faults, called **fault zones**.

The effects of earthquakes depend on how easily the rocks move when energy builds up. There is always friction between rock surfaces. Rough and jagged surfaces produce more friction than smooth ones. More energy needs to build up to move rough surfaces against each other, so more energy is released once they start to move. This creates a more severe earthquake.

Earthquake damage. A new fault line now runs under the house, where the rocks have broken.

Scientists measure the strength of earthquakes using the Richter scale. The higher the number on the scale, the more energy is released. Big earthquakes are rare. But small earthquakes are extremely common, especially those so small that only sensitive instruments can detect them.

Characteristic effects of shallow shocks in populated areas	Richter scale	Number of earthquakes per year
Almost total damage	more than 8.0	0.1–0.2
Great damage	more than 7.4	4
Serious damage, rails bent	7.0–7.3	15
Considerable damage to buildings	6.2–6.9	100
Slight damage to buildings	5.5–6.1	500
Felt by all	4.9–5.4	1400
Felt by many	4.3–4.8	4800
Felt by some	3.5–4.2	30 000
Not felt but recorded	2.0–3.4	800 000

Some scientists have suggested earthquake damage could be reduced by lubricating the rock surfaces. As an experiment, they drilled down to a fault and pumped in water. This reduced the friction and there was a minor earthquake. When the water was pumped away from the fault, the rocks were stopped from moving for a long time. But when the rocks did move again, they caused a severe earthquake.

Questions

1 Look at the map on the opposite page. Is there any link between regions which have active volcanoes and regions where earthquakes are likely to occur? Explain your answer.

2 Where in the oceans are volcanoes and earthquakes most likely to occur?

3 What is a meant by:
   a the *epicentre* of an earthquake?
   b a *fault*?

4 Why might pumping water into a fault help prevent an earthquake? What problems might this treatment cause?

14.4 Faults and folds

Earthquakes happen when rocks move along fault lines. Many regions of the world have experienced huge earthquakes in the distant past, even though they are free of them today. Britain is one example.

Fault zones

Britain is not in a major earthquake zone. However, the large number of fault zones running through the rocks of Britain provide evidence suggesting that, in the distant past, its rocks were affected by major earthquakes.

Some British fault zones are very large, running for many kilometres across the country and continuing deep into the Earth.

A fault through layers of sedimentary rock.

Sometimes, faults have occurred where rocks have been pushed together (**compression**). In other places, they have occurred where rocks have been pulled apart (**tension**). The fault shown above was caused by tension.

Some of the main landscape features in Britain are caused by fault zones. For example, the Midland Valley in Scotland is a block of fairly soft rocks that has dropped down between two fault zones. The rocks to the north are the hard metamorphic rocks of the Highlands. To the south are the hard sandstones of the Southern Uplands. These features are illustrated in the diagram below.

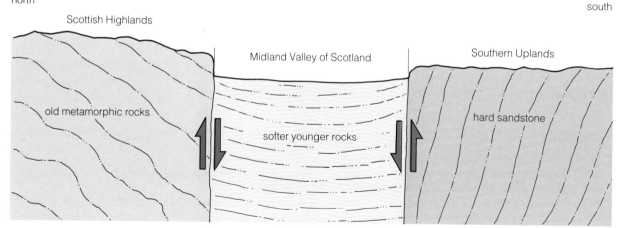

Folds

Faults are formed when large forces build up quickly in rocks. The rocks behave in a brittle way and break. If the force builds up very slowly and continues for a long time, the rock may bend instead. This is how **folds** are formed.

Folds are evidence that rocks can behave in a *plastic* (flexible) way. They show that rocks have been *compressed*. Folds can be very severe. Huge forces would have been needed to produce the folds shown in the photograph on the right.

Upfolds are called **anticlines**. Downfolds are called **synclines**.

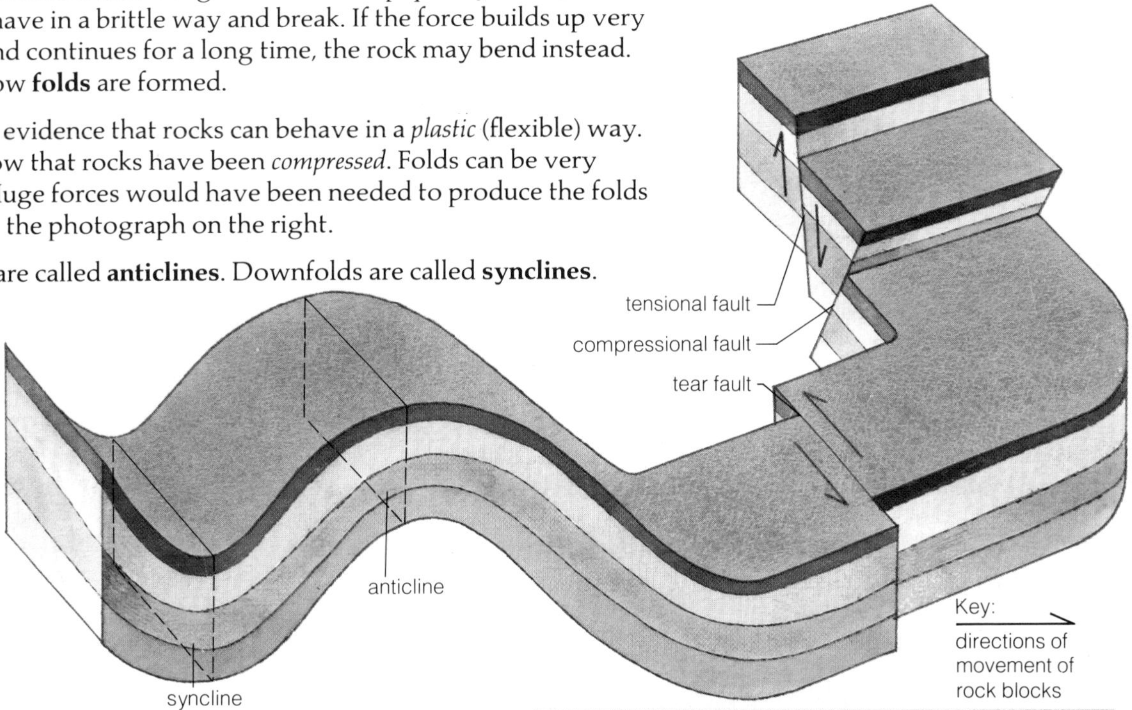

tensional fault

compressional fault

tear fault

anticline

syncline

Key:

directions of movement of rock blocks

Some very large folds make major landscape features. The south of England is made of a set of sedimentary layers which have folded up and have then been eroded. A typical section through the landscape is shown in the diagram below.

Folded rocks at Lulworth Cove in Dorset.

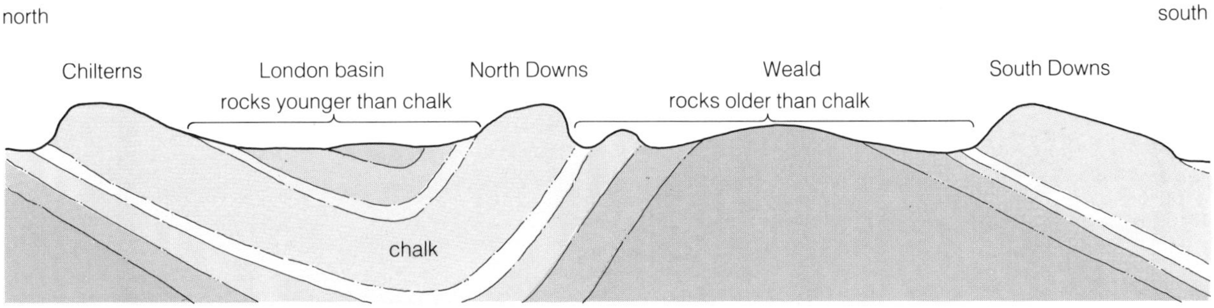

north   south

Chilterns   London basin   North Downs   Weald   South Downs
          rocks younger than chalk   rocks older than chalk

chalk

Questions

1 What evidence is there that Britain once experienced huge earthquakes?
2 Look at the photograph on the opposite page. Was this fault caused by tension or compression? What clues are there in the picture to support your answer?

3 a Copy the diagram above. Using dots, draw in the layers which would have been above the landscape before erosion took place.
  b Find and label a *syncline* on your diagram.
  c Find and label an *anticline* on your diagram.

14.5 Metamorphic rocks

The rocks in the photograph are not just folded. From their layering, they might appear to be sedimentary rocks. In fact they are **metamorphic** rocks. The word 'metamorphic' means 'changed'. Scientists are interested in knowing what the rocks have been changed from and what they have been changed by.

Most metamorphic rocks began as sedimentary rocks — especially the very common clay-type rocks such as **shale**. They are made when heat, pressure, or both, affect existing rocks. The banding in metamorphic rocks looks like sedimentary layers. But whereas sedimentary rocks have *fragments* which are *layered*, metamorphic rocks have *crystals* which are *lined up*.

Slate is a metamorphic rock formed from shale. It has tiny, flat crystals which can only be seen with a microscope. Slate splits parallel to its flat crystals. The splitting is called **cleavage**. **Schist** and **gneiss** are also metamorphic rocks formed from clay-type rocks. However, they are produced under more extreme conditions than slate. Gneiss is the most altered rock. It is the **highest grade** of metamorphic rock.

Metamorphic rocks.

Regional metamorphism

Slate, schist, and gneiss are often found in areas of fold mountains, for example the Scottish Highlands. These mountains are the eroded remains of a mountain chain which was as high as the Alps are today.

Metamorphic rocks made inside fold mountains by both heat and pressure are called **regional metamorphic rocks**. The graph on the next page shows how, as conditions become more extreme, clay-type rocks such as shale are changed into slate, then schist, then gneiss. The most extreme conditions melt the rock, forming magma which may eventually become new igneous rock.

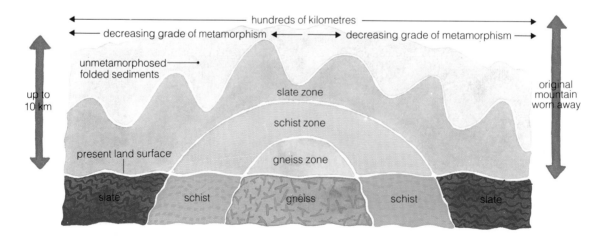
A fold mountain belt.

Thermal metamorphism

Metamorphic rocks can also be caused by heat alone. The heat usually comes from an igneous rock that is cooling down underground. This type of change is called **thermal metamorphism**.

There is an example of thermal metamorphism in the Whin Sill described on page 233. The sill is a thick layer of once-molten igneous rock, dolerite. Above and below the sill are limestone layers which have been changed by heat into a crumbly form of **marble** where they touched the hot igneous rock. This is called a **baked margin**.

The bigger the source of heat, the larger the effects will be. The area of metamorphic rocks around a large igneous mass is called the **metamorphic aureole**.

Thermal metamorphism does not change shale to slate, schist, or gneiss. Instead, it changes shale to **spotted rocks** and **hornfels**.

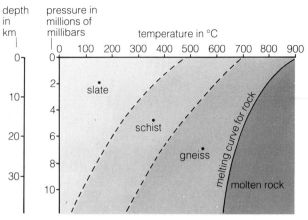

The conditions responsible for regional metamorphism (see section on previous page).

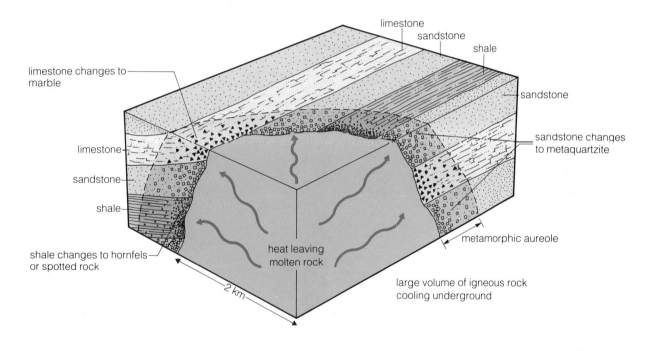

Thermal metamorphic rocks around a large granite pluton.

Questions

1 Gneiss is a *higher grade* of metamorphic rock than schist. What does this mean?
2 Where are *regional metamorphic rocks* formed?
3 In the diagram on the left, why is gneiss, and not schist or slate, formed deep in the fold mountains?

4 From the graph on the top of this page:
  a What is the pressure at a depth of 10 km?
  b What is the minimum pressure needed to turn shale into schist, if the rock temperature is 200 °C?
5 What is thermal metamorphism? Give an example.

14.6 Drifting continents

The theory of continental drift

Early in the seventeenth century, Francis Bacon noted that newly-mapped American coasts looked as though they would fit into the coasts of Europe and Africa, like pieces of a giant jigsaw.

As rocks and fossils were mapped on both sides of the Atlantic, more evidence appeared to suggest that the continents had once been joined up. There was also evidence that land areas had drifted through different climatic zones. For example, in Britain, rocks of Carboniferous age, over 300 million years old, appeared to have formed in tropical swamps near the Equator. And 100 million years later Britain was covered by deserts, suggesting that it was in a similar position to the Sahara today.

In 1915, Alfred Wegener put forward his theory of **continental drift**. He suggested that, millions of years ago, all the continents had once been joined in a giant supercontinent which he named **Pangaea**. This had split and the parts had moved around the world.

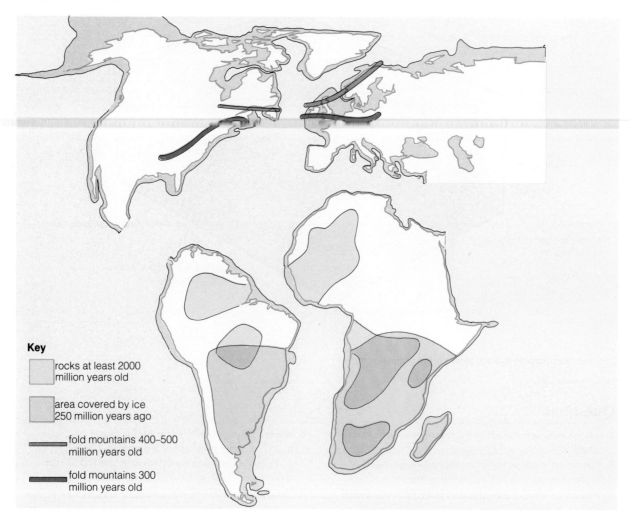

Key

☐ rocks at least 2000 million years old

☐ area covered by ice 250 million years ago

— fold mountains 400–500 million years old

— fold mountains 300 million years old

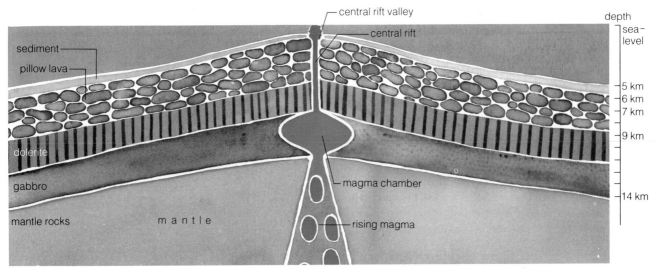

central rift valley
central rift
depth
sea-level
sediment
pillow lava
5 km
6 km
7 km
9 km
dolerite
magma chamber
gabbro
14 km
mantle rocks — m a n t l e
rising magma

Cross-section of an ocean ridge.

Sea-floor spreading

The continents are part of the Earth's outer layer, called the **crust**. Scientists think that they 'float' on the denser, hotter, and more plastic rocks of the **mantle** beneath.

Scientists have found that, on ocean ridges, the layer of sediment is far thinner than elsewhere. Further observations have shown that the ocean floor at ridges is made of new rock, formed by volcanic activity. This effect is called **sea-floor spreading**.

The lavas are pushed out of vents, rather like toothpaste is if you give the tube a series of jerky squeezes. The rocks are basalt lavas. They are called **pillow lavas** because of their rounded shape. Beneath them are rocks formed from magma which solidifies before reaching the surface.

As the lavas are slowly pushed out to form new rock, the crust under the ocean gets bigger and the continents move apart.

Pillow lava, exposed during road-cutting in the Balkans. The lava originally formed underwater.

Questions

1 What evidence is there that, in the distant past, the continents may not have been in their present positions on the Earth?

2 Study the map on the opposite page. Describe three pieces of evidence which suggest that the continents may once have been joined together.

3 In the diagram at the top of the page:
   a Where are the oldest pillow lavas?
   b Why does the layer of sediment get thicker as you move further out from the ocean ridge?

4 Pillow lavas have smaller crystals than the rocks beneath them. Suggest reasons for this.

14.7 Colliding continents

Subduction zones

If extra ocean floor is being produced at ocean ridges, it seems hard to explain why the Earth is not getting bigger. However, in the 1950s, Hugo Benioff discovered a region of earthquake activity which ran at an angle down through the Earth's surface from the sea floor. This is now explained as slabs of ocean floor moving down into the Earth at an angle. This process is called **subduction**.

Some examples of **subduction zones** are shown in the diagrams on the right. The moving slabs are huge pieces of crust (and upper mantle) called **plates**. There is more on plates in the next Unit.

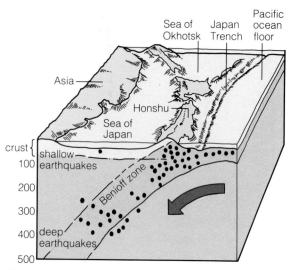

The pattern of earthquakes below Japan

Many subduction zones are found around the edge of the Pacific Ocean. Some are not obvious because the **ocean trench** that usually marks the top of the zone may be filled with sediment.

Making volcanoes

As the slab of ocean floor moves down, the friction produces heat. Hot fluids rise from the descending slab and melt the rocks above, forming magma. This is less dense and more gassy than its surroundings, so it forces its way upwards and erupts as a volcano.

Volcanoes at ocean trenches are more explosive than the vents on ocean ridges. These different volcanic environments produce the two different types of volcano, **explosive** and **quiet** (see page 234).

If the subduction zone is in an oceanic area, the volcanoes will build islands. These make curved lines called **island arcs**.

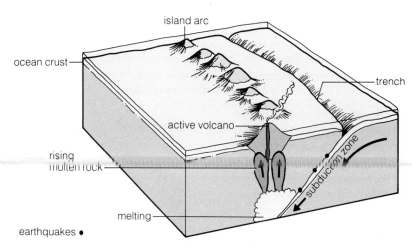

A subduction zone where ocean plate meets ocean plate

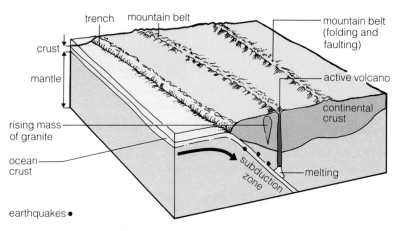

A subduction zone where ocean plate meets continental plate

The diagram shows four stages of mountain formation:

Top left: volcanoes, sediment in bottom of trench, sediments, continent, continent

Top right: sediment from trench and top of oceanic crust, sediment worn off continent, continent, continent

Bottom left: metamorphism, granite melt, wedge of new sediment eroded from mountains, continent, continent

Bottom right: wedge of new sediment eroded from mountains, continent, continent

How mountain ranges form.

Continental collisions

If subduction at plate edges is faster than the spreading at the mid-ocean ridge, then the ocean will shrink. The continents on either side of the ocean will be dragged together. Eventually they will meet, and the island arcs will be trapped between them.

The results of continental collisions are **fold mountains**. For example, the Himalayas in Nepal, north of India, are made of rocks pushed up when India (once a separate land mass) collided with Asia. The Alps are the result of Africa colliding with Europe.

Continents are built up of many ancient subduction zones and sediments. Island arcs collided with other island arcs, and made the first continents. As oceans opened and closed, blocks of continental crust grew larger and thicker. Because continental crust is fairly low in density, it tends not to be subducted.

Ancient chains of fold mountains have been eroded so that now we see only the deeply eroded roots. These are made of folded and faulted metamorphic rocks. Most of the Scottish Highlands are the remains of material formed long ago, at the edges of oceans.

Explosive volcanoes, like this one in the Andes mountains of South America, mostly erupt andesite. The rock is named after the mountain range.

Questions

1 What is a *subduction zone*?
2 Why do volcanoes form near subduction zones?
3 Why can metamorphic rocks be found near subduction zones?

4 How are ocean trenches produced?
5 How are island arcs produced?
6 Why do some continents have fold mountains between them? Give an example of some fold mountains.

14.8 The theory of plates

To explain continental drift, scientists have developed the theory of **plate tectonics**. According to this theory, the Earth's outer shell is made up of the huge **plates** shown on the map below. Their edges are marked by earthquakes, volcanoes, or both. The plates move at different speeds and in different directions. For example, laser surveys show that Africa and America are separating by about 4 cm per year. Where the crust is widening, as along the Mid-Atlantic Ridge, **rift valleys** form (see the diagram on page 241).

Plate boundaries

The lines where one plate moves against another are called **plate boundaries**. There are various types:

Constructive boundaries   These occur at ocean ridges, where new ocean floor is being formed. Here, plates are getting bigger: they are being constructed.

Destructive boundaries   These happen at subduction zones. As the denser plate slides under the less dense one, it gets smaller. In other words, the subducted plate is being destroyed.

Conservative boundaries   These are where plates slide past each other so that neither gets bigger or smaller. In other words, both plates are being *conserved*.

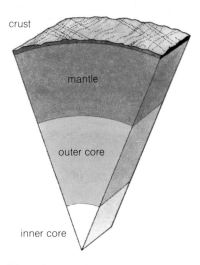

What scientists think the Earth's structure is like. Plates are made up of the crust and a small part of the mantle beneath it.

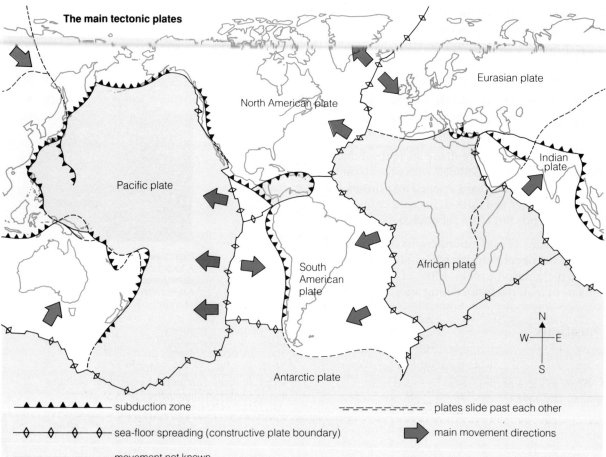

The main tectonic plates

Eurasian plate

North American plate

Indian plate

Pacific plate

African plate

South American plate

Antarctic plate

▲▲▲▲▲▲ subduction zone

◇ ◇ ◇ ◇ sea-floor spreading (constructive plate boundary)

- - - - - movement not known

- - - - - plates slide past each other

➡ main movement directions

Data from inside the Earth

To explain how plates can slide over the inner part of the Earth, scientists need information about the Earth's structure. They get this by studying **seismic waves**, the shock waves sent out by earthquakes. Two types of seismic wave, called **P-waves** and **S-waves**, travel deep into the Earth, and can be detected by sensitive instruments when they reach the surface, often many thousands of kilometres away. Seismic wave speeds are affected by conditions inside the Earth. By comparing the arrival times of the waves, scientists can deduce what these conditions are like.

In the chart on the right, the **geotherm** is a line showing the temperature at different depths. The other temperature line shows the melting point of rock. Deeper rocks are under more pressure from the rocks above, which raises their melting point.

The speeds of seismic waves depend on the density of the rock. At about 10 km down, the speeds of P and S-waves suddenly rise. Here, the lower-density crust meets the mantle beneath it.

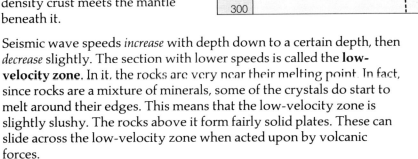

Seismic wave speeds *increase* with depth down to a certain depth, then *decrease* slightly. The section with lower speeds is called the **low-velocity zone**. In it, the rocks are very near their melting point. In fact, since rocks are a mixture of minerals, some of the crystals do start to melt around their edges. This means that the low-velocity zone is slightly slushy. The rocks above it form fairly solid plates. These can slide across the low-velocity zone when acted upon by volcanic forces.

The theory of plate tectonics is the best explanation so far for the movements of the Earth's crust. As new evidence emerges, this may change.

Questions

1 Iceland lies on the Mid-Atlantic Ridge. It is about 500 km across. How long has it taken to erupt this material from the ridge?
2 Which type of plate boundary is found in each of these places? Use the map on the left to answer.
  a The Mid-Atlantic Ridge.
  b Along the western edge of South America.
  c Along the western edge of California.

Use the chart above to answer the following:
3 What is the temperature in the Earth at a depth of:
  a 100 km? b 200 km? c 300 km?
4 At what depth is the melting point 1500 °C?
5 At what depth does the crust meet the mantle?
6 At what depth is the top of the low-velocity zone?
7 Why are the rocks found in the low-velocity zone slightly slushy?

14.9 Recycling rocks

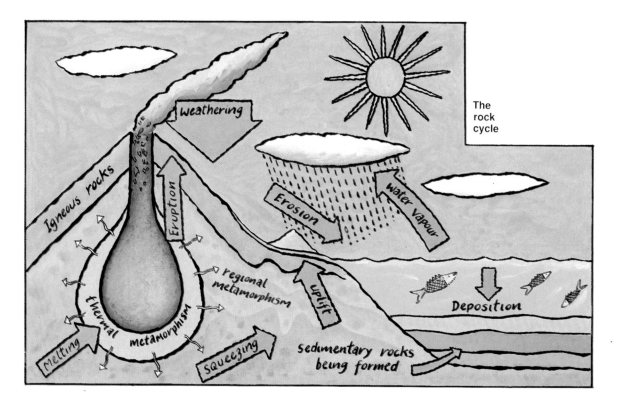

The rocks on the Earth's surface are made of materials that have been recycled by natural processes. Those processes are still making new rocks. This recycling of the Earth's outer layers is called the **rock cycle**.

Some parts of the rock cycle happen on the Earth's surface. They are driven by gravity and by energy from the Sun. The Sun's heat causes the air movements which we call weather. It also drives the water cycle.

Some parts of the rock cycle happen inside the Earth. The energy driving these internal processes comes from the heat released by the radioactive decay of certain atoms.

The rock cycle concentrates economically important minerals. For example, there are rich deposits of gold, copper, and tin near some plate boundaries. They occur because hot, high-pressure sea water dissolves minerals when it moves down through cracks deep in the crust. Later, the minerals are deposited when the water rises up through cracks on the ocean floor. These cracks are called **veins**.

Questions

Use the diagram above to help you answer the following:

1 How do sedimentary rocks turn into metamorphic rocks?
2 How do igneous rocks turn into sedimentary rocks?
3 How do metamorphic rocks turn into igneous rocks?
4 Some minerals were deposited by water as it passed through cracks in the ocean floor. Explain why, today, those minerals might be found in hills on dry land.

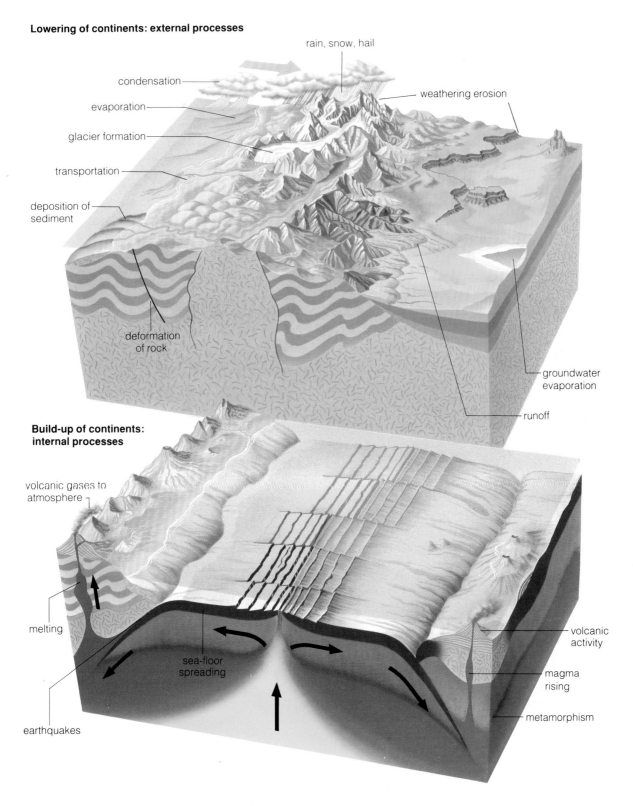

Lowering of continents: external processes

rain, snow, hail

condensation

evaporation

weathering erosion

glacier formation

transportation

deposition of sediment

deformation of rock

groundwater evaporation

runoff

Build-up of continents: internal processes

volcanic gases to atmosphere

melting

sea-floor spreading

earthquakes

volcanic activity

magma rising

metamorphism

A summary of Earth processes Internal processes build up the surface, and erosion wears it down.

Questions on Section 14

1 The diagram below shows a section through a volcano. The three main types of rock, *igneous*, *sedimentary*, and *metamorphic*, are all to be found in the area.

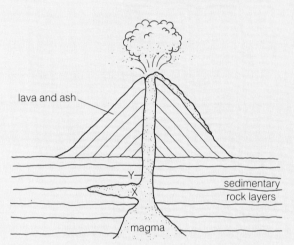

a Which type of rock is the lava?
b If the hot, liquid magma at X cools and solidifies, which type of rock will it become?
c Which type of rock might you expect to find at Y? Give reasons for your answer.
d The volcano in the diagram has a high slope angle. Why do some volcanoes have a much lower slope angle than this?
e Explain what these terms mean when used to describe volcanoes:
   i   active
   ii   quiet
   iii dormant

2 The diagrams below show two igneous rocks in close-up:

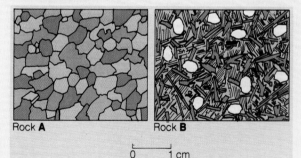

a Which of the two igneous rocks cooled most quickly? Give reasons for your answer.
b What caused the round holes in rock B?
c Which of the two rocks is most likely to have solidified underground? Give reasons for your answer.

3 Give reasons for each of the following:
a Earthquakes and volcanoes tend to be concentrated in the same narrow zones of the Earth's crust.
b Some of the continents seem to fit together like the pieces of a giant jigsaw.
c In some parts of the world's oceans, there are long ridges; in others there are long trenches.
d Near an ocean ridge, the sea-floor sediment is thinner.

4 This map of Scotland shows how a major fault has broken an area of granite:

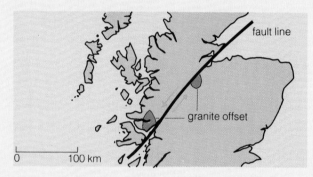

a Measure the displacement of the fault. Use the scale to work out the displacement in kilometres.
b What kind of fault is this: *tensional*, *compressional*, or *tear*?
c Explain how it is possible for large masses of solid granite to have moved so far.

5 a What is meant by a *metamorphic* rock?
b If one metamorphic rock is of a *higher grade* than another, what does this mean?
c Put these metamorphic rocks in order of grade, starting with the lowest grade: *schist, slate, gneiss*.
d Copy and complete these statements by writing in either *regional* or *thermal* in each case:
   i   .... metamorphism only affects small areas.
   ii   .... metamorphism is caused by heat and pressure.
   iii The roots of fold mountain chains are mainly made of .... metamorphic rocks.

6 Draw labelled diagrams to show each of the following:
a A pluton.
b An anticline and a syncline.
c The formation of new crust at an ocean ridge.
d Subduction.
e The formation of fold mountains by continental collision.

7 These diagrams show the boundaries between tectonic plates.

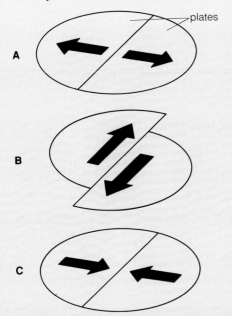

a Which one shows a *destructive* boundary?
b Which one shows a *constructive* boundary?
c Which one shows a *conservative* boundary?
d On which one will there be an ocean ridge?
e At which one will subduction occur?
f Near which one might fold mountains be formed?

8 The diagram below shows the rock cycle. Copy and complete the diagram by filling in the boxes, using the labels provided:

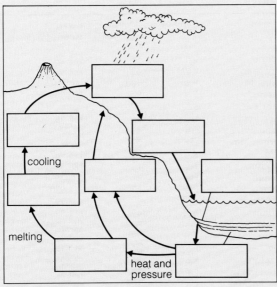

Labels for boxes

igneous rocks	erosion and transport	weathering
sedimentary rocks	deposition	magma
metamorphic rocks	uplift and folding	

9 Boreholes were drilled to collect sediment from just above the pillow lavas under the South Atlantic Ocean. These are the results:

Distance from ocean ridge in km	Age of sediment in million years (from fossil evidence)
190	9.5
450	22
790	39
1440	73

a Plot the results on a graph.
b Draw in the best-fit straight line.
c On average, how much ocean floor is being made per year on this side of the ridge?
d How much is the ocean floor widening per century?
e One specimen has been stored unlabelled. Fossils date it at about 51 million years old. How far from the ridge was it probably collected?

10 When an earthquake occurs, two different types of seismic waves (P-waves and S-waves) travel through the Earth at different speeds. The graph below shows the travel times of the waves at places of increasing distance from the focus of the earthquake:

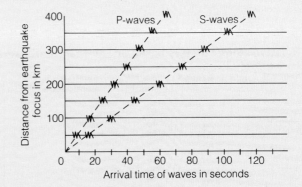

a At a distance of 400 km from the focus, what is the time difference between the P-waves arriving and the S-waves arriving?
b Which waves arrive first, P-waves or S-waves?
c How long does it take P-waves to travel 400 km?
d From your answer to part **c**, find the speed of P-waves.
e How long does it take S-waves to tr   l 400 km?
f From your answer to part **e**, find the speed of S-waves.
g At a monitoring station, the time dif   ence between P-waves arriving and S-waves arriving, is 32 seconds. How far is the station from the focus of the earthquake?

The periodic table and atomic masses

The periodic table showing Group 1 and 2, the transition metals, and the lanthanide and actinide series.

Group

| Period | Group 1 | Group 2 | | Transition metals | | | | | | | | | | | 1 | $^1_1$H hydrogen |

Period 1: $^1_1$H hydrogen

Period 2: $^7_3$Li lithium, $^9_4$Be beryllium

Period 3: $^{23}_{11}$Na sodium, $^{24}_{12}$Mg magnesium

The transition metals

Period 4: $^{39}_{19}$K potassium, $^{40}_{20}$Ca calcium, $^{45}_{21}$Sc scandium, $^{48}_{22}$Ti titanium, $^{51}_{23}$V vanadium, $^{52}_{24}$Cr chromium, $^{55}_{25}$Mn manganese, $^{56}_{26}$Fe iron, $^{59}_{27}$Co cobalt, $^{59}_{28}$Ni nickel, $^{64}_{29}$Cu copper, $^{65}_{30}$Zn zinc

Period 5: $^{85}_{37}$Rb rubidium, $^{88}_{38}$Sr strontium, $^{89}_{39}$Y yttrium, $^{91}_{40}$Zr zirconium, $^{93}_{41}$Nb niobium, $^{96}_{42}$Mo molybdenum, $^{98}_{43}$Tc technetium, $^{101}_{44}$Ru ruthenium, $^{103}_{45}$Rh rhodium, $^{106}_{46}$Pd palladium, $^{108}_{47}$Ag silver, $^{112}_{48}$Cd cadmium

Period 6: $^{133}_{55}$Cs caesium, $^{137}_{56}$Ba barium, $^{139}_{57}$La lanthanum, $^{178.5}_{72}$Hf hafnium, $^{181}_{73}$Ta tantalum, $^{184}_{74}$W tungsten, $^{186}_{75}$Re rhenium, $^{190}_{76}$Os osmium, $^{192}_{77}$Ir iridium, $^{195}_{78}$Pt platinum, $^{197}_{79}$Au gold, $^{201}_{80}$Hg mercury

Period 7: $^{223}_{37}$Fr francium, $^{226}_{88}$Ra radium, $^{227}_{89}$Ac actinium

Lanthanides: $^{140}_{58}$Ce cerium, $^{141}_{59}$Pr prae-sodimium, $^{144}_{60}$Nd neodimium, $^{147}_{61}$Pm promethium, $^{150}_{62}$Sm samarium, $^{152}_{63}$Eu europium, $^{157}_{64}$Gd gadolinium, $^{159}_{65}$Tb terbium

Actinides: $^{232}_{90}$Th thorium, $^{231}_{91}$Pa prot-actinium, $^{238}_{92}$U uranium, $^{237}_{93}$Np neptunium, $^{242}_{94}$Pu plutonium, $^{243}_{95}$Am americium, $^{247}_{96}$Cm curium, $^{247}_{97}$Bk berkelium

Relative atomic masses based on internationally agreed figures

Element	Symbol	Atomic number	Relative atomic mass	Element	Symbol	Atomic number	Relative atomic mass
Actinium	Ac	89		Erbium	Er	68	167.26
Aluminium	Al	13	26.9815	Europium	Eu	63	151.96
Americium	Am	95		Fermium	Fm	100	
Antimony	Sb	51	121.75	Fluorine	F	9	18.9984
Argon	Ar	18	39.948	Francium	Fr	87	
Arsenic	As	33	74.9216	Gadolinium	Gd	64	157.25
Astatine	At	85		Gallium	Ga	31	69.72
Barium	Ba	56	137.34	Germanium	Ge	32	72.59
Berkelium	Bk	97		Gold	Au	79	196.967
Beryllium	Be	4	9.0122	Hafnium	Hf	72	178.49
Bismuth	Bi	83	208.980	Helium	He	2	4.0026
Boron	B	5	10.811	Holmium	Ho	67	164.930
Bromine	Br	35	79.909	Hydrogen	H	1	1.00797
Cadmium	Cd	48	112.40	Indium	In	49	114.82
Caesium	Cs	55	132.905	Iodine	I	53	126.9044
Calcium	Ca	20	40.08	Iridium	Ir	77	192.2
Californium	Cf	98		Iron	Fe	26	55.847
Carbon	C	6	12.01115	Krypton	Kr	36	83.80
Cerium	Ce	58	140.12	Lanthanum	La	57	138.91
Chlorine	Cl	17	35.453	Lawrencium	Lw	103	
Chromium	Cr	24	51.996	Lead	Pb	82	207.19
Cobalt	Co	27	58.9332	Lithium	Li	3	6.939
Copper	Cu	29	63.54	Lutetium	Lu	71	174.97
Curium	Cm	96		Magnesium	Mg	12	24.312
Dysprosium	Dy	66	162.50	Manganese	Mn	25	54.9380
Einsteinium	Es	99		Mendelevium	Md	101	

Group

3	4	5	6	7	0
					$^{4}_{2}$He helium
$^{11}_{5}$B boron	$^{12}_{6}$C carbon	$^{14}_{7}$N nitrogen	$^{16}_{8}$O oxygen	$^{19}_{9}$F fluorine	$^{20}_{10}$Ne neon
$^{27}_{13}$Al aluminium	$^{28}_{14}$Si silicon	$^{31}_{15}$P phosphorus	$^{32}_{16}$S sulphur	$^{35.5}_{17}$Cl chlorine	$^{40}_{18}$Ar argon
$^{70}_{31}$Ga gallium	$^{73}_{32}$Ge germanium	$^{75}_{33}$As arsenic	$^{79}_{34}$Se selenium	$^{80}_{35}$Br bromine	$^{84}_{36}$Kr krypton
$^{115}_{49}$In indium	$^{119}_{50}$Sn tin	$^{122}_{51}$Sb antimony	$^{128}_{52}$Te tellurium	$^{127}_{53}$I iodine	$^{131}_{54}$Xe xenon
$^{204}_{81}$Tl thallium	$^{207}_{82}$Pb lead	$^{209}_{83}$Bi bismuth	$^{210}_{84}$Po polonium	$^{210}_{85}$At astatine	$^{222}_{86}$Rn radon

$^{162}_{66}$Dy dysprosium	$^{165}_{67}$Ho holmium	$^{167}_{68}$Er erbium	$^{169}_{69}$Tm thulium	$^{173}_{70}$Yb ytterbium	$^{175}_{71}$Lu lutecium
$^{251}_{98}$Cf californium	$^{254}_{99}$Es einsteinium	$^{253}_{100}$Fm fermium	$^{256}_{101}$Md mendelevium	$^{254}_{102}$No nobelium	$^{257}_{103}$Lr lawrencium

Approximate atomic masses for calculations

Element	Symbol	Atomic mass for calculations
Aluminium	Al	27
Bromine	Br	80
Calcium	Ca	40
Carbon	C	12
Chlorine	Cl	35.5
Copper	Cu	64
Helium	He	4
Hydrogen	H	1
Iodine	I	127
Iron	Fe	56
Lead	Pb	207
Lithium	Li	7
Magnesium	Mg	24
Manganese	Mn	55
Neon	Ne	20
Nitrogen	N	14
Oxygen	O	16
Phosphorus	P	31
Potassium	K	39
Silicon	Si	28
Silver	Ag	108
Sodium	Na	23
Sulphur	S	32
Zinc	Zn	65

Element	Symbol	Atomic number	Relative atomic mass
Mercury	Hg	80	200.59
Molybdenum	Mo	42	95.94
Neodymium	Nd	60	144.24
Neon	Ne	10	20.179
Neptunium	Np	93	
Nickel	Ni	28	58.71
Niobium	Nb	41	92.906
Nitrogen	N	7	14.0067
Nobelium	No	102	
Osmium	Os	76	190.2
Oxygen	O	8	15.9994
Palladium	Pd	46	106.4
Phosphorus	P	15	30.9738
Platinum	Pt	78	195.09
Plutonium	Pu	94	
Polonium	Po	84	
Potassium	K	19	39.102
Praseodymium	Pr	59	140.907
Promethium	Pm	61	
Protactinium	Pa	91	
Radium	Ra	88	
Radon	Rn	86	
Rhenium	Re	75	186.2
Rhodium	Rh	45	102.905
Rubidium	Rb	37	85.47
Ruthenium	Ru	44	101.07

Element	Symbol	Atomic number	Relative atomic mass
Samarium	Sm	62	150.35
Scandium	Sc	21	44.956
Selenium	Se	34	78.96
Silicon	Si	14	28.086
Silver	Ag	47	107.868
Sodium	Na	11	22.9898
Strontium	Sr	38	87.62
Sulphur	S	16	32.064
Tantalum	Ta	73	180.948
Technetium	Tc	43	
Tellurium	Te	52	127.60
Terbium	Tb	65	158.924
Thallium	Tl	81	204.37
Thorium	Th	90	232.038
Thulium	Tm	69	168.934
Tin	Sn	50	118.69
Titanium	Ti	22	47.90
Tungsten	W	74	183.85
Uranium	U	92	238.03
Vanadium	V	23	50.942
Xenon	Xe	54	131.30
Ytterbium	Yb	70	173.04
Yttrium	Y	39	88.905
Zinc	Zn	30	65.37
Zirconium	Zr	40	91.22

Answers to numerical questions

page 45   10 a The correct values for rubidium are: melting point 39 °C, boiling point 688 °C

page 73   4 58   **5 a** 32   **b** 254   **c** 16   **d** 46   **e** 132

page 75   4 a 1 g   **b** 127 g   **c** 35.5 g   **d** 71 g
5 a 32 g   **b** 64 g   **6** 138 g   **7 a** 9 moles   **b** 3 moles

page 77   1 sulphur 50%, oxygen 50%   **2** hydrogen 5%, oxygen 60%   **3** hydrogen 11.1%   **4** 0.1 moles per dm³, or 0.1 M

page 79   1 a 1   **b**   4 g   **3** FeS   **4** H_2S

page 80   1 a 64 g   **b** 48 g   **c** 48 g   **d** 60 g   **e** 355 g
f 1.4 g   **g** 4 g   **h** 0.6 g   **i** 24 g   **2 a** 2 g   **b** 4 g   **c** 32 g
d 35.5 g   **e** 248 g   **f** 1024 g   **3 a** 1 mole   **b** 2 moles
c 1 mole   **d** 2 moles   **e** 0.2 moles   **f** 0.1 moles   **g** 0.4 moles   **h** 0.2 moles   **i** 2 moles   **j** 0.05 moles   **4 a** 80 g of sulphur   **b** 80 g of oxygen   **c** 8 moles of chlorine atoms
d 1 mole of oxygen molecules   **e** 4 moles of sulphur atoms   **5 a** 1 litre of 2 M sodium chloride   **b** 1 litre of 1 M sodium chloride   **c** 100 cm³ of 2 M sodium chloride
d 1 litre of 1 M sodium hydroxide   **e** 40 cm³ of 1 M sodium chloride   **6 a** 18 g   **b** 90 g   **c** 160 g   **d** 250 g   **e** 34 g
f 48 g   **g** 30 g   **h** 8 g   **i** 15.8 g   **j** 325 g   **8** The missing numbers are: **a** 40, 16, 1, 56   **b** 1.6, 5.6   **c** 0.16   **d** 6, 3
e 71.4%   **9 a** There is 80% copper in each sample

page 81   10 a 64 g   **b** 4 moles   **c** 2 moles   **d** MnO_2
11 a 2.4 g   **b** 0.1 moles   **c** 1.6 g   **d** 0.1 moles   **e** MgO
12 a 106.5   **b** 3 moles   **c** 1 mole   **d** $AlCl_3$ **e** 1 M
f 0.1 M   **13 c** NH_3, 17 g; CO_2, 44 g; H_2, 2 g; O_2, 32 g
d 2 moles   **e** 6 moles   **f** 22 g   **14 c** The missing figures are: group 4, 0.19 g; group 5, 0.20 g   **e** 0.16 g   **f** 16 g
g 2 moles   **h** 1 mole   **i** Cu_2O

page 87   2 b 2 moles   **c i** 32 g   **ii** 8 g   **3 b** $CuCO_3$, 124 g; CuO, 80 g; CO_2, 44 g   **c i** 11 g   **ii** 0.25   **iii** 6 dm³

page 96   4 a 217 g   **b** 20.1 g of mercury, 1.6 g of oxygen
5 a 3 g   **b** sulphur   **c** 11 g of iron(II) sulphide and 6 g of sulphur   **7 a** 233 g   **b** 0.1 moles   **c** 0.1 moles of each
d 1 litre (or 1 dm³)

page 97   8 a 48 dm³ 48 000 cm³   **b** 12 dm³ 12 000 cm³
c 2.4 dm³ 2400 cm³   **d** 7.2 dm³ 7200 cm³   **9 b i** 2 moles
ii 168 g   **c i** 1 mole   **ii** 24 dm³ or 24 000 cm³   **d i** 12 dm³
or 12 000 cm³   **ii** 1.2 dm³ or 1200 cm³   **10 b** 0.5 moles
c i 28 g   **ii** 22 g   **iii** 12 dm³ or 12 000 cm³   **11 b i** 18 g
ii 1.67 moles per dm³ or 1.67 M   **iii** 0.83 moles   **iv** 20 dm³
or 20 000 cm³   **12 a** 12 °C   **c i** 50.4 kJ   **ii** 2217.6 kJ
13 b i 30 kJ   **ii** 280.4 kJ   **14 c** 55.6 kJ

page 105   1 a 3600 coulombs   **b** 0.0373   **c** 1.194 g

page 111   11 b 3   **c** 201 451 g or 201.45 kg
12 c i 0.2 A   **ii** 0.119 g   **d** Fe^{3+} **13 b ii** 2 moles of electrons   **iii** 193 000   **c i** 0.298 dm³ or 298 cm³

page 115   3 a i 29 cm³   **ii** 39 cm³   **b i** 0.65 minutes
ii 1.5 minutes   **c i** 5 cm³ of hydrogen per minute   **ii** 2 cm³ of hydrogen per minute

page 117   1 a i 60 cm³   **ii** 60 cm³

page 119   1 a experiment 1, 0.55 g; experiment 2, 0.95 g
b experiment 1, 0.55 g; experiment 2, 0.95 g   **c** experiment 1, g per minute; experiment 2, g per minute

page 126   2 c i 14 cm³   **ii** 9 cm³   **iii** 8 cm³   **iv** 7 cm³
v 2 cm³   **d i** 14 cm³ per minute   **ii** 9 cm³ per minute
iii 8 cm³ per minute   **iv** 7 cm³ per minute   **v** 2 cm³ per minute   **e** 40 cm³   **f** 5 minutes   **g** 8 cm³ per minute

page 127   7 i 0.5 g   **10 e** 3 moles

page 221   5 a 0.1 m/s   **b** 10 mm   **c** 0.001 m/s

page 225   2 0.065 cm

page 228   4 c 33 cm³   **d** oxygen 33%, nitrogen 67%
e oxygen 21%, nitrogen 78%

page 231   4 a 22 °, 14 °

page 239   4 a 3 million mb   **b** 6 million mb

page 245   1 12 500 000 years   **3 a** 1400 °C   **b** 1550 °C
c 1600 °C   **4** 150 km   **5** 10 km   **6** 65 km

pages 248–249   4 a 100 km   **9 c** 2 cm   **d** 4 m
e 1020 km   **10 a** 53 s   **c** 65 s   **d** 6.2 m/s   **e** 118 s
f 3.4 m/s   **g** 230 km

Index

If more than one page number is given, you should look up the **bold** one first.